Coles Notes
Study Guide
Physics

WILEY

John Wiley & Sons Canada, Ltd.

Coles Notes Study Guide Physics

Published by:
John Wiley & Sons Canada, Ltd.
6045 Freemont Blvd.
Mississauga, Ontario
L5R 4J3

Note: If you purchased this book without a cover, you should be aware that this book is stolen property. It was reported as "unsold and destroyed" to the publisher, and neither the author nor the publisher has received any payment for this "stripped book."

ISBN: 978-1-118-30726-7

Printed in Canada

1 2 3 4 5 WEB 16 15 14 13 12

For general information on our other products and services or to obtain technical support, please contact our Customer Care Department within the U.S. at 877-762-2974, outside the U.S. at 317-572-3993, or fax 317-572-4002.

Wiley also publishes its books in a variety of electronic formats. Some content that appears in print may not be available in electronic books. For more information about Wiley products, please visit our website at www.wiley.com.

WILEY

FSC www.fsc.org

MIX
Paper from responsible sources
FSC® C004071

Table of Contents

Introduction

Physics is a branch of physical science that deals with physical changes of objects. The mental, idealized models on which it is based are most frequently expressed in mathematical equations that simplify the condition of the real world for ease of analysis. Even though the equations are derived from ideal conditions, they approximate real situations closely enough to allow accurate prediction of the behaviours of complex systems.

The primary task in studying physics is to understand its basic principles. Understanding these formal principles enables better understanding of the phenomena observed in the universe.

The system of units used throughout this book is the International System of Units (SI). This is the metric system, with which you may be familiar. The basic units are length (metre, m), mass (kilogram, kg), time (second, s), temperature (degrees celsius,°C, or Kelvin, K), electric current (amperes, A), and amount (mole). Standard prefixes are often used; for example, millimetre (10^{-3} metre) is abbreviated mm.

This book is written with a broad audience in mind. Therefore, concepts are presented at varying levels of mathematical sophistication. Each topic is generally first presented in a manner that requires only basic geometry and trigonometry. In some cases, formulae requiring knowledge of calculus are given for those readers familiar with it. However, calculus is not required to understand the concepts in this book.

Why You Need This Book

Can you answer yes to any of the following questions?

- Do you need to review the fundamentals of physics quickly?
- Do you need a course supplement to physics?
- Do you need to prepare for your physics test?
- Do you need a concise reference for physics?

If so, then *Coles Notes Study Guide Physics* is for you!

How to Use This Book

You can use this book in any way that fits your personal style for study and review — you decide what works best with your needs. You can either read the book from cover to cover or just look for the information you want and put it back on the shelf for later. Here are a few recommended ways to search for topics.

- Look for areas of interest in the book's table of contents.
- Flip through the book, looking for subject areas at the top of each page.
- Get a glimpse of what you'll gain from a chapter by reading through the Chapter Check-In at the beginning of each chapter.
- Use the Chapter Checkout at the end of each chapter to gauge your grasp of the important information you need to know.
- Test your knowledge more completely in the Review.

- Use the glossary to find key terms fast. This book defines new terms and concepts where they first appear in the chapter. If a word is boldfaced, you can find a more complete definition in the book's glossary.
- Or flip through the book until you find what you're looking for—we organized this book to gradually build on key concepts.

Chapter 1
CLASSICAL MECHANICS

CHAPTER CHECK-IN

- ❑ Understanding motion (kinematics) in one and two dimensions, and rotational motion
- ❑ Applying Newton's Laws and analyzing force diagrams
- ❑ Using the concepts of energy and momentum
- ❑ Learning about periodic motion and elasticity
- ❑ Applying classical mechanics to fluids

Mechanics is the study of the motion of material objects. Classical or Newtonian mechanics deals with objects and motions familiar in our everyday world. Most people possess some intuition about classical mechanics; we all have watched a ball fly through the air or a bicycle tire spin. You should not be afraid to connect the formalism in this book with your intuition. Indeed, this is often the easiest way to see the answer to a difficult problem. Allow the formal physics and math to illuminate what you already know.

Although textbook examples usually deal with blocks, springs, and other mundane devices, keep in mind that classical mechanics describes phenomena on a vast range of scales, from large molecules such as DNA to the planets of our solar system and beyond. In this chapter, you learn about the basic laws that these diverse systems all obey.

Kinematics in One Dimension

Kinematics analyzes the positions and motions of objects as a function of time, without regard to the causes of motion. It involves the relationships between the quantities **displacement** *(d)*, **velocity** *(v)*, **acceleration** *(a)*, and **time** *(t)*. The first three of these quantities are vectors.

Definition of a vector

A **vector** is a physical quantity with direction as well as magnitude, for example, velocity or force. In contrast, a quantity that has only magnitude and no direction, such as temperature or time, is called a **scalar.** A vector is commonly denoted by an arrow drawn with a length proportional to the given magnitude of the physical quantity and with direction shown by the orientation of the head of the arrow.

Displacement and velocity

Imagine that a car begins traveling along a road after starting from a specific sign post. To know the exact position of the car after it has traveled a given distance, it is necessary to know not only the kilometres it traveled but also its heading. The **displacement**, defined as the change in position of the object, is a vector with the magnitude as a distance, such as 10 kilometres, and a direction, such as east. Velocity is a vector expression with a magnitude equal to the speed traveled and with an indicated direction of motion. For motion defined on a number line, the direction is specified by a positive or negative sign. Average velocity is mathematically defined as

$$\text{average velocity} = \frac{\text{total displacement}}{\text{time elapsed}}$$

Note that displacement (distance from starting position) is *not* the same as distance traveled. If a car travels one kilometre east and then returns one kilometre west, to the same position, the total displacement is zero and so is the average velocity over this time period. Displacement is measured in units of length, such as metres or kilometres, and velocity is measured in units of length per time, such as metres/second (metres per second).

Average acceleration

Acceleration, defined as the rate of change of velocity, is given by the following equation:

$$\text{average acceleration} = \frac{\text{final velocity} - \text{initial velocity}}{\text{time elapsed}}$$

Acceleration units are expressed as length per time divided by time such as metres/second/second or in abbreviated form as m/s^2.

Graphical interpretations of displacement, velocity, and acceleration

The distance versus time graph in Figure 1-1 shows the progress of a person: (I) standing still, (II) walking with a constant velocity, and (III) walking with a slower constant velocity. The slope of the line yields the speed. For example, the speed in segment II is

$$\frac{(4-0)\,\text{m}}{(10-5)\,\text{s}} = \frac{4\,\text{m}}{5\,\text{s}} = .8\,\text{m/s}$$

Figure 1-1 Motion of a walking person.

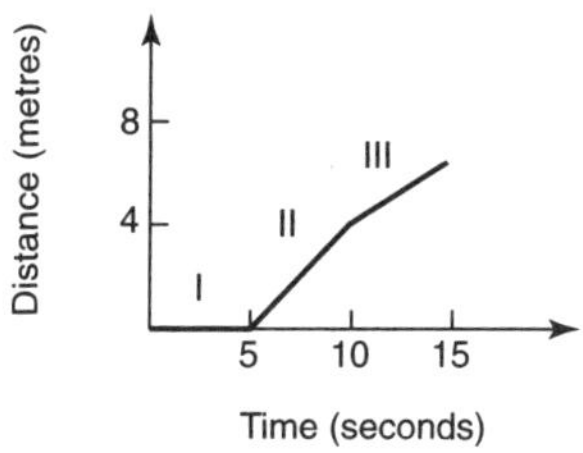

Each segment in the velocity versus time graph in Figure 1-2 depicts a different motion of a bicycle: (I) increasing velocity, (II) constant velocity, (III) decreasing velocity, and (IV) velocity in a direction opposite the initial direction (negative). The area between the curve and the time axis represents the distance traveled. For example, the distance traveled during segment I is equal to the area of the triangle with height 15 and base 10. Because the area of a triangle is (1/2)(base)(height), then (1/2)(15 m/s)(10 s) = 75 m. The magnitude of acceleration equals the calculated slope. The acceleration calculation for segment III is (–15 m/s)/(10 s) = –1.5 m/s/s or –1.5 m/s^2.

The more realistic distance-versus-time curve in Figure 1-3(a) illustrates gradual changes in the motion of a moving car. The speed is nearly constant in the first 2 seconds, as can be seen by the nearly constant slope of the line; however, between 2 and 4 seconds, the speed is steadily decreasing and the **instantaneous velocity** describes how fast the object is moving at a given instant.

Figure 1-2 Accelerating motion of a bicycle.

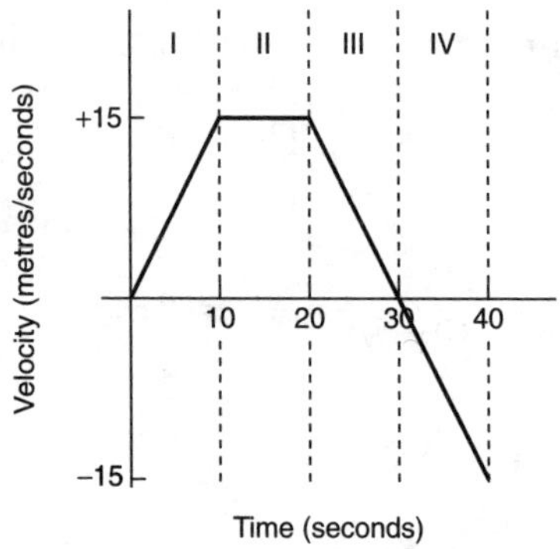

Figure 1-3 Motion of a car: (a) distance, (b) velocity, and (c) acceleration change in time.

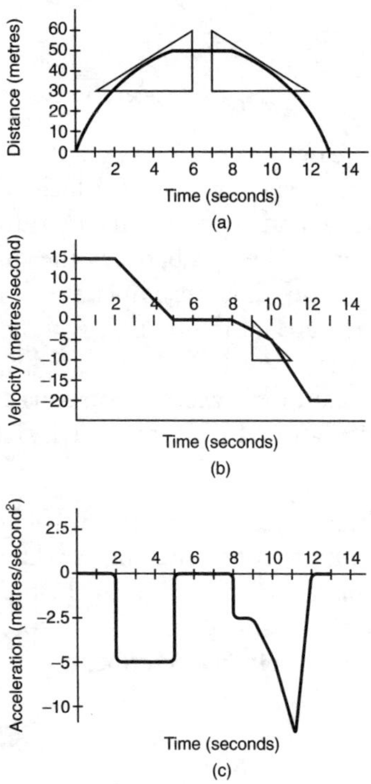

Instantaneous velocity can be read on an odometer in the car. It is calculated from a graph as the slope of a tangent to the curve at the specified time. The slope of the line sketched at 4 seconds is 6 m/s. Figure 1-3(b) is a sketch of the velocity-versus-time graph constructed from the slopes of the distance-versus-time curve. In like fashion, the **instantaneous acceleration** is found from the slope of a tangent to the velocity-versus-time curve at a given time. The instantaneous acceleration-versus-time graph in Figure 1-3(c) is the sketch of the slopes of the velocity-versus-time graph of Figure 1-3(b). With the vertical arrangement shown, it is easy to compute the displacement, velocity, and acceleration of a moving object at the same time.

For example, at time $t = 10$ s, the displacement is 47 m, the velocity is –5 m/s, and the acceleration is $-5\ \text{m/s}^2$.

Definitions of instantaneous velocity and instantaneous acceleration

The instantaneous velocity, by definition, is the limit of the average velocity as the measured time interval is made smaller and smaller. In formal terms, $v = \lim_{\Delta t \to 0} \Delta d / \Delta t$. The notation $\lim_{\Delta t \to 0}$ means the ratio $\Delta d / \Delta t$ is evaluated as the time interval approaches zero. Similarly, instantaneous acceleration is defined as the limit of the average acceleration as the time interval becomes infinitesimally short. That is, $a = \lim_{\Delta t \to 0} \Delta v / \Delta t$.

Motion with constant acceleration

When an object moves with constant acceleration, the velocity increases or decreases at the same rate throughout the motion. The average acceleration equals the instantaneous acceleration when the acceleration is constant. A negative acceleration can indicate either of two conditions:

- **Case 1:** The object has a decreasing velocity in the positive direction.
- **Case 2:** The object has an increasing velocity in the negative direction.

For example, a ball tossed up will be under the influence of a negative (downward) acceleration due to gravity. Its velocity will decrease while it travels upward (case 1); then, after reaching its highest point, the velocity will increase downward as the object returns to earth (case 2).

Using v_o (velocity at the beginning of time elapsed), v_f (velocity at the end of the time elapsed), and t for time, the constant acceleration is

$$a = \frac{v_f - v_o}{t} \text{ or } v_f = v_o + at \qquad \text{[Equation 1]}$$

Substituting the average velocity as the arithmetic average of the original and final velocities $v_{avg} = (v_o + v_f)/2$ into the relationship between distance and average velocity $d = (v_{avg})(t)$ yields

$$d = \frac{1}{2}(v_o + v_f)t \qquad \text{[Equation 2]}$$

Substitute v_f from Equation 1 into Equation 2 to obtain

$$d = v_o t + \frac{1}{2}at^2 \qquad \text{[Equation 3]}$$

Finally, substitute the value of t from Equation 1 into Equation 2 for

$$v_f^2 = v_o^2 + 2ad \qquad \text{[Equation 4]}$$

These four equations relate v_o, v_f, t, a, and d. Note that each equation has a different set of four of these five quantities. Table 1-1 summarizes the equations for motion in a straight line under constant acceleration.

A special case of constant acceleration occurs for an object under the influence of gravity. If an object is thrown vertically upward or dropped, the acceleration due to gravity of -9.8 m/s^2 is substituted in the above equations to find the relationships among velocity, distance, and time.

Table 1-1 Equations and Variables of Kinematics in One Dimension

Equation	*Information Given by Equation*	v_o	v_f	t	a	d
$v_f = v_o + at$	Velocity as a function of time	✓	✓	✓	✓	X
$d = \frac{1}{2}(v_o + v_f)t$	Displacement varying with velocity and time	✓	✓	✓	X	✓
$d = v_o t + \frac{1}{2}at^2$	Displacement as a function of time	✓	X	✓	✓	✓
$v_f^2 = v_o^2 + 2ad$	Velocity as a function of displacement	✓	✓	X	✓	✓

Kinematics in Two Dimensions

Up to this time, only forward and backward motion along a number line has been considered; however, our world is three-dimensional. For easier analysis, many motions can be simplified to two dimensions. For example, an object fired into the air moves in a vertical, two-dimensional plane; also, horizontal motion over the earth's surface is two-dimensional for short distances. Elementary vector algebra is required to examine the relationships between vector quantities in two dimensions.

Addition and subtraction of vectors: geometric method

The vector **A** shown in Figure 1-4(a) represents a velocity of 10 m/s northeast, and vector **B** represents a velocity of 20 m/s at 30 degrees north of east. (A vector is named with a letter in **boldface,** nonitalic type, and its magnitude is named with the same letter in regular, *italic* type. You will often see vectors in the figures of the book that are represented by their magnitudes in the mathematical expressions.) Vectors may be moved over the plane if the represented length and direction are preserved.

Figure 1-4 Graphical addition of vectors, **A** + **B** = **C**.

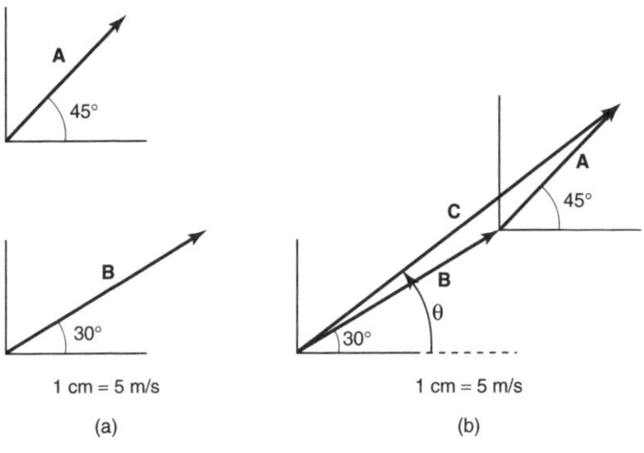

In Figure 1-4(b), the same vectors are positioned to be geometrically added. The tail of one vector, in this case **A**, is moved to the head of the other vector (**B**). The vector sum (**C**) is the vector that extends from the tail of one vector to the head of the other. To find the magnitude of **C,** measure along its length and use the given scale to determine the velocity represented. To find the direction θ of **C,** measure the angle to the horizontal axis at the tail of **C.**

Figure 1-5(a) shows that **A** + **B** = **B** + **A.** The sum of the vectors is called the **resultant** and is the diagonal of a parallelogram with sides **A** and **B.** Figure 1-5(b) illustrates the construction for adding four vectors. The resultant vector is the vector that results in the one that completes the polygon.

Figure 1-5 (a) A + B = B + A. (b) Graphical addition of several vectors.

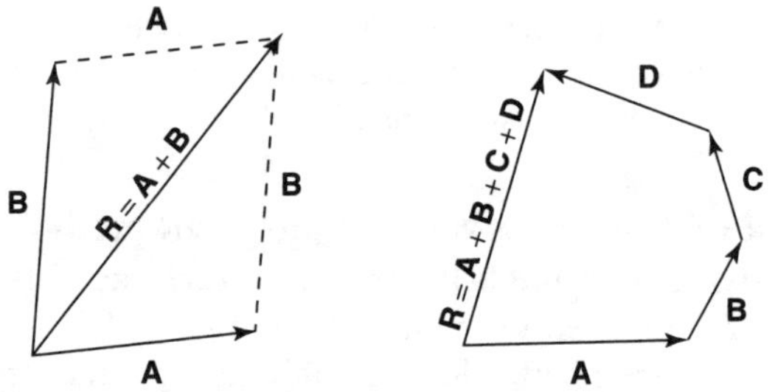

(a) (b)

To subtract vectors, place the tails together. The difference of the two vectors (**D**) is the vector that begins at the head of the subtracted vector (**B**) and goes to the head of the other vector (**A**). An alternate method is to add the negative of a vector, which is a vector with the same length but pointing in the opposite direction. The second method is demonstrated in Figure 1-6.

Figure 1-6 Graphical subtraction of vectors, A – B = D.

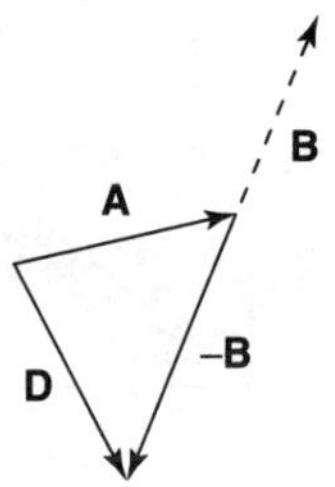

Addition and subtraction of vectors: Component method

For precision in adding vectors, an analytical method using basic trigonometry is required because scale drawings do not give accurate values.

Consider vector **A** in the rectangular coordinate system of Figure 1-7. The vector **A** can be expressed as the sum of two vectors along the *x* and *y* axes, $\mathbf{A} = \mathbf{A}_x + \mathbf{A}_y$, where $\mathbf{A}_x$ and $\mathbf{A}_y$ are called the **components** of **A**. The direction of $\mathbf{A}_x$ is parallel to the *x* axis, and that of $\mathbf{A}_y$ is parallel to the *y* axis.

The magnitudes of the components are obtained from the definitions of the sine and cosine of an angle: $\cos\theta = A_x/A$ and $\sin\theta = A_y/A$, or

$$A_x = A\cos\theta$$
$$A_y = A\sin\theta$$

Figure 1-7 Components of a vector.

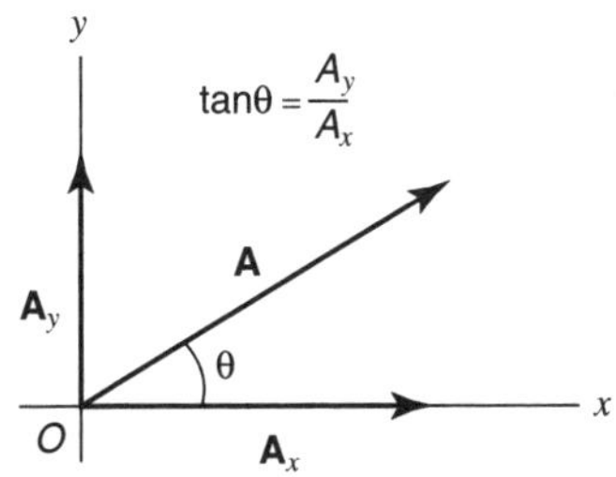

To add vectors numerically, first find the components of all the vectors. The signs of the components are the same as the signs of the cosine and sine in the given quadrant. Then, sum the components in the *x* direction, and sum the components in the *y* direction. As shown in Figure 1-8, the sum of the *x* components and the sum of the *y* components of the given vectors (**A** and **B**) comprise the *x* and *y* components of the resultant vector (**C**).

These resultant components form the two sides of a right angle with a hypotenuse of the magnitude of **C**; thus, the magnitude of the resultant is

$$C = \sqrt{C_x^2 + C_y^2}$$

Figure 1-8 Component method of vector addition, **A** + **B** = **C**.

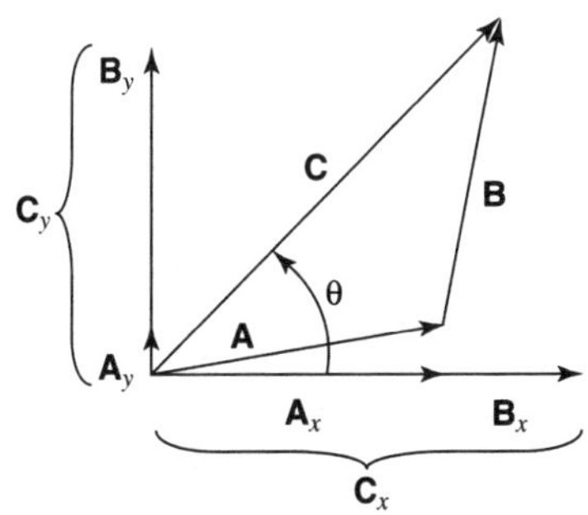

The direction of the resultant (**C**) is calculated from the tangent because $\tan\theta = C_x / C_y$. To solve for the angle θ, use $\theta = \tan^{-1}(C_y / C_x)$.

The procedure can be summarized as follows:

1. **Sketch the vectors on a coordinate system.**
2. **Find the *x* and *y* components of all the vectors, with the appropriate signs.**
3. **Sum the components in both the *x* and *y* directions.**
4. **Find the magnitude of the resultant vector from the Pythagorean theorem.**
5. **Find the direction of the resultant vector using the tangent function.**

Follow the same procedure to subtract vectors by calculating the appropriate algebraic sum of the components in Step 3.

Multiplication of vectors

The dot product: There are two different ways in which two vectors may be multiplied together. The first is the dot product, also called the scalar product, which is written **A** · **B**. This can be evaluated in two ways:

1 $\mathbf{A}\cdot\mathbf{B} = A_x B_x + A_y B_y$

- $\mathbf{A}\cdot\mathbf{B} = AB\cos\theta$, where θ is the angle between the vectors when they are set tail to tail, and *A* and *B* are the lengths of the vectors.

Note that the order of the vectors does not matter and that the result of the dot product is a scalar rather than a vector. Note that if two vectors are perpendicular, their dot product is zero according to the second rule above.

Cross product: The second way to multiply vectors is called the **cross product** or the **vector product**. It is written **A** × **B**. It can be evaluated in two ways:

- $\mathbf{A}\times\mathbf{B} = (A_x B_y - A_y B_x)\mathbf{z}$, when the vectors **A** and **B** both are in *x-y* plane. The **z** indicates that the result is a vector that points along the *z* axis. In general, the vector resulting from a cross product is always perpendicular to both of the vectors being multiplied together.
- $\mathbf{A}\times\mathbf{B} = AB\mathbf{z}\sin\theta$, where θ is the angle between the vectors **A** and **B** when they are placed tail to tail. Again, the result is a vector perpendicular to **A** and **B** (and therefore points along the *z* axis if **A** and **B** are in the *x-y* plane).

The result of a cross product does depend on the order of the vectors. Note from the first rule that **A** × **B** = –**B** × **A.** Also, if **A** and **B** are parallel, the second rule implies that their cross product is zero.

Finally, the cross product give rise to the "right hand rule," which allows you to easily determine the direction of the resulting vector. For the general expression **A** × **B** = **C**, point your thumb in the direction of **A.** Now point your index finger in the direction of **B**; if necessary, flip over your hand. The vector **C** points outward from your palm. For an illustration of this procedure, flip ahead to Chapter 4 (Figure 4-19), where the rule is applied to the equation **F** = q**v** × **B**.

Velocity and acceleration vectors in two dimensions

For motion in two dimensions, the earlier kinematics equations must be expressed in vector form. For example, the average velocity vector is $\boldsymbol{v} = (\boldsymbol{d}_f - \boldsymbol{d}_o)/t$, where $\boldsymbol{d}_o$ and $\boldsymbol{d}_f$ are the initial and final displacement vectors and t is the time elapsed. As noted earlier, the velocity and displacement vectors are shown in bold type, whereas the scalar *(t)* is not. In similar fashion, the average acceleration vector is $\boldsymbol{a} = (\boldsymbol{v}_f - \boldsymbol{v}_o)/t$, where $\boldsymbol{v}_o$ and $\boldsymbol{v}_f$ are the initial and final velocity vectors.

An important point is that the acceleration can arise from a change in the magnitude of the velocity (speed) as well as from a change in the direction of the velocity. If an object travels around a circle at a constant speed, there is an acceleration due to the change in the direction of the velocity, even though the magnitude of the velocity does not change. A mass moves in a horizontal circle with a constant speed in Figure 1-9. The velocity vectors at positions 1 and 2 are subtracted to find the average acceleration, which is directed toward the centre of the circle. (Note that the average acceleration vector is placed at the midpoint of the path in the given time interval.)

Figure 1-9 Velocity and acceleration vectors of an object moving in a circle.

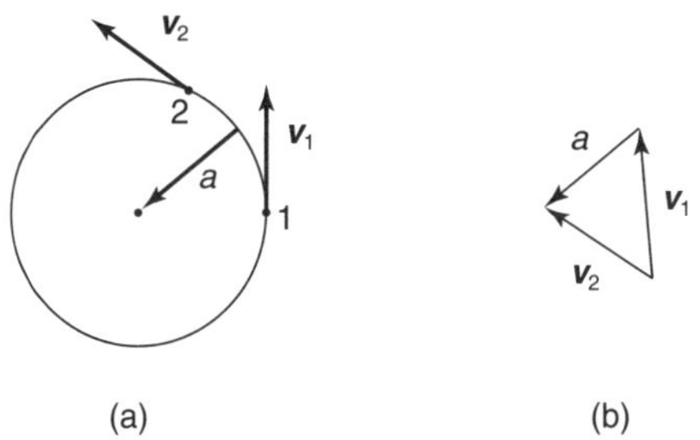

The following discussion summarizes the four different cases for acceleration in a plane:

- **Case 1:** Zero acceleration
- **Case 2:** Acceleration due to changing direction but not speed
- **Case 3:** Acceleration due to changing speed but not direction
- **Case 4:** Acceleration due to changing both speed and direction.

Imagine a ball rolling on a horizontal surface that is illuminated by a stroboscopic light. Figure 1-10(a) shows the position of the ball at even intervals of time along a dotted path. Case 1 is illustrated in positions 1 through 3; the magnitude and direction of the velocity do not change (the pictures are evenly spaced and in a straight line), and therefore, there is no acceleration. Case 2 is indicated for positions 3 through 5; the ball has constant speed but changing direction, and therefore, an acceleration exists. Figure 1-10(b) illustrates the subtraction of $\mathbf{v}_3$ and $\mathbf{v}_4$ and the resulting acceleration toward the centre of the arc. Case 3 occurs from positions 5 to 7; the direction of the velocity is constant, but the magnitude changes. The acceleration for this portion of the path is along the direction of motion. The ball curves from position 7 to 9, showing case 4; the velocity changes both direction and magnitude. In this case, the acceleration is directed nearly upward between 7 and 8 and has a component toward the centre of the arc due to the change in direction of the velocity and a component along the path due to the change in the magnitude of the velocity.

Figure 1-10 (a) Path of a ball on a table. (b) Acceleration between points 3 and 4.

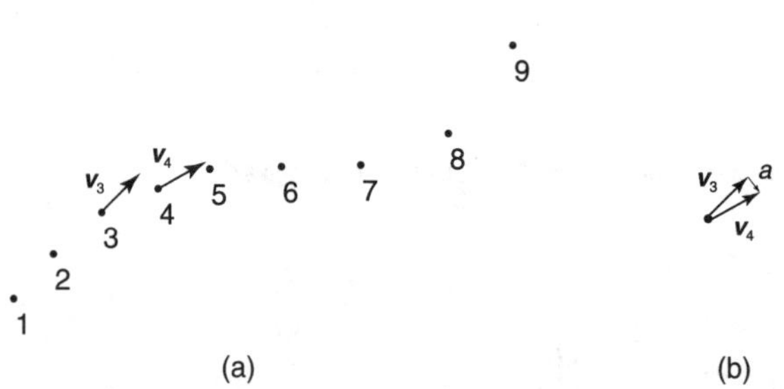

Projectile motion

Anyone who has observed a tossed object—for example, a baseball in flight—has observed **projectile motion**. To analyze this common type of motion, three basic assumptions are made: (1) acceleration due to gravity is constant and directed downward, (2) the effect of air resistance is negligible, and (3) the surface of the earth is a stationary plane (that is, the curvature of the earth's surface and the rotation of the earth are negligible).

To analyze the motion, separate the two-dimensional motion into vertical and horizontal components. Vertically, the object undergoes constant acceleration due to gravity. Horizontally, the object experiences no acceleration and, therefore, maintains a constant velocity. This velocity is illustrated in Figure 1-11 where the velocity components change in the *y* direction; however, they are all of the same length in the *x* direction (constant). Note that the velocity vector changes with time due to the fact that the vertical component is changing.

Figure 1-11 Projectile motion.

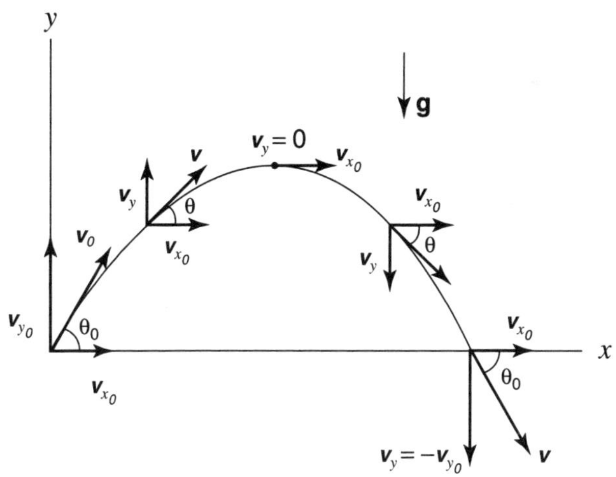

In this example, the particle leaves the origin with an initial velocity (v_o), up at an angle of θ_o. The original *x* and *y* components of the velocity are given by $v_{xo} = v_o \cos \theta_o$ and $v_{yo} = v_o \sin \theta_o$.

With the motions separated into components, the quantities in the x and y directions can be analyzed with the one-dimensional motion equations subscripted for each direction: for the horizontal direction, $v_x = v_{xo}$ and $x = v_{xo}t$; for vertical direction, $v_y = v_{yo} - gt$ and $y = v_{yo}t - (1/2)gt^2$, where x and y represent distances in the horizontal and vertical directions, respectively, and the acceleration due to gravity *(g)* is 9.8 m/s^2. (The negative sign is already incorporated into the equations.) If the object is fired down at an angle, the y component of the initial velocity is negative. The speed of the projectile at any instant can be calculated from the components at that time from the Pythagorean theorem, and the direction can be found from the inverse tangent on the ratios of the components:

$$v = \sqrt{v_x^2 + v_y^2}$$

$$\theta = \tan^{-1}\left|v_y / v_x\right|$$

Other information is useful in solving projectile problems. Consider the example shown in Figure 1-11 where the projectile is fired up at an angle from ground level and returns to the same level. The time for the projectile to reach the ground from its highest point is equal to the time of fall for a freely falling object that falls straight down from the same height. This equality of time is because the horizontal component of the initial velocity of the projectile affects how far the projectile travels horizontally but not the time of flight. Projectile paths are parabolic and, therefore, symmetric. Also for this case, the object reaches the top of its rise in half of the total time *(T)* of flight. At the top of the rise, the vertical velocity is zero. (The acceleration is always g, even at the top of the flight.) These facts can be used to derive the **range** of the projectile, or the distance traveled horizontally. At maximum height, $v_y = 0$ and $t = T/2$; therefore, the velocity equation in the vertical direction becomes $0 = v_o \sin\theta - g\,T/2$ or solving for T, $T = (2v_o \sin\theta)/g$.

Substitution into the horizontal distance equation yields $R = (v_o \cos\theta)\,T$. Substitute T in the range equation and use the trigonometry identity $\sin 2\theta = 2\sin\theta\cos\theta$ to obtain an expression for the range in terms of the initial speed and angle of motion, $R = (v_o^2/g)\sin 2\theta$. As indicated by this expression, the maximum range occurs when $\theta = 45$ degrees because, at this value of θ, $\sin 2\theta$ has its maximum value of 1. Figure 1-12 sketches the trajectories of projectiles thrown with the same initial speed at differing angles of inclination.

Figure 1-12 Range of projectiles launched at different angles.

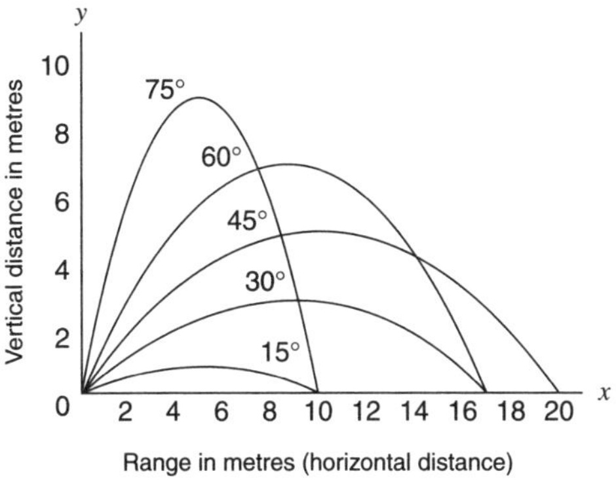

Uniform circular motion

For uniform motion of an object in a horizontal circle of radius *(R)*, the constant speed is given by $v = 2\pi R/T$, which is the distance of one revolution divided by the time for one revolution. The time for one revolution *(T)* is defined as **period.** During one rotation, the head of the velocity vector traces a circle of circumference $2\pi v$ in one period; thus, the magnitude of the acceleration is $a = 2\pi v/T$. Combine these two equations to obtain two additional relationships in other variables: $a = v^2/R$ and $a = (4\pi^2/T^2)\,R$.

The displacement vector is directed out from the centre of the circle of motion. The velocity vector is tangent to the path. The acceleration vector directed to the centre of the circle is called **centripetal acceleration.** Figure 1-13 shows the displacement, velocity, and acceleration vectors at different positions as the mass travels in a circle on a frictionless horizontal plane.

Figure 1-13 Uniform circular motion.

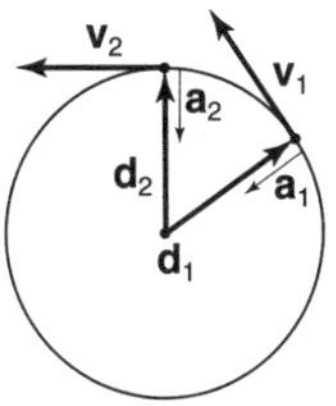

Dynamics

The study of dynamics goes beyond the relationships between the variables of motion as illuminated in kinematics to the cause of motion, which is force.

Newton's laws of motion

Newton's first law of motion, also called the law on inertia, states that an object continues in its state of rest or of uniform motion unless compelled to change that state by an external force. The law appears to contain two separate statements. The first statement—that a state of rest will continue unless a force is applied—seems intuitively correct. The second statement—that an object will continue with a constant velocity unless compelled to change by an impressed force—seems contrary to common experience. It is important to realize that objects observed to slow down are being compelled to change by a frictional force. **Friction** is a retarding force that is ever present in our everyday world. For the ideal—the absence of outside forces acting on the object, as described by the law—friction must be eliminated. The value of the law is the introduction of the concept of **force** as a push or pull that causes a body to change its state of motion.

Newton's second law of motion states that if a net force acts on an object, it will cause an acceleration of that object. The law addresses the cause and effect relationship between force and motion commonly stated as $F = m\mathbf{a}$, where m is the proportionality constant (mass). Force is measured in SI units of **newtons,** abbreviated N.

Newton's third law of motion states that for every action there is an equal and opposite reaction. Therefore, if one object exerts a force on a second object, the second exerts an equal and oppositely directed force on the first one.

Mass and weight

Mass and **weight** are distinctly different physical quantities, a fact that cannot be emphasized too strongly. Mass is the property that lends an object a reluctance to change its state of motion. Mass is the measure of the amount of matter in an object. Masses are compared on an equal-arm balance. If a loaded two-pan balance is level on earth, it will be level in a different gravitational field, as for example, on the moon. Thus, mass is an invariant quantity; it is measured in units of kilograms. A mass of 1 kilogram will experience an acceleration of 1 m/s^2 under the action of a force of 1 newton.

The force that the earth exerts on an object of specific mass is called the object's weight on earth. Weight is a force measured in units of newtons and is a vector quantity. The expression for weight is **W** = m**g**, where **g** is the acceleration due to gravity. A spring scale translates the force of attraction between an object and the earth into a reading of weight. In contrast to a measurement of mass, weight is not an invariant. An object on a spring scale on earth would not weigh the same on the moon because the pull of gravity on the object differs in the two locations.

Force diagrams

To better understand the relationship between force and acceleration in a particular case, it is helpful to use a **force diagram,** also called a **free-body diagram.** An object that is not moving is said to be in a **static equilibrium.** An example is a weight hanging by two ropes from the ceiling (see Figure 1-14). To analyze this problem, consider the forces acting on the knot joining the ropes. Then,

1. **Make the force diagram.**
2. **Find the components of forces not directed along the coordinate axes and write the force equation for each axis.**
3. **Solve the simultaneous equations for the tensions.**

The following examples illustrate these procedures. (All of the following vector diagrams are drawn to scale.)

Example 1: What is the tension in each rope in Figure 1-14?

Figure 1-14 Force diagram of a suspended weight.

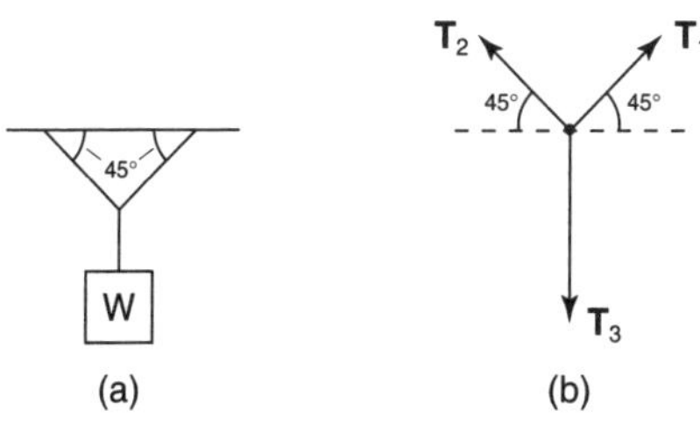

1. **Make the force diagram.**

   The tension in the lower rope attached to the mass must be mg directed downward; therefore, $T_3 = -mg$.

2. **Find the components of forces not directed along the coordinate axes and write the force equation for each axis.**

   Components of the forces in the *x* direction are

$$T_2 \cos 45° - T_1 \cos 45° = 0$$

   Components of the forces in the y direction are

$$T_2 \sin 45° + T_1 \sin 45° - mg = 0$$

3. **Solve the simultaneous equations for the tensions.**

Solution: $T_1 = T_2 = \frac{mg}{\sqrt{2}}$

To minimize computation errors, show the components on a separate force diagram as shown in Figure 1-14.

Example 2: Now, see if you can make the free-body diagram and set up the force equations for a pail on the end of a rope that is accelerating upward. Find the tension in the rope and the acceleration of the pail. See Figure 1-15(a).

Figure 1-15 Force diagram of a bucket being lifted.

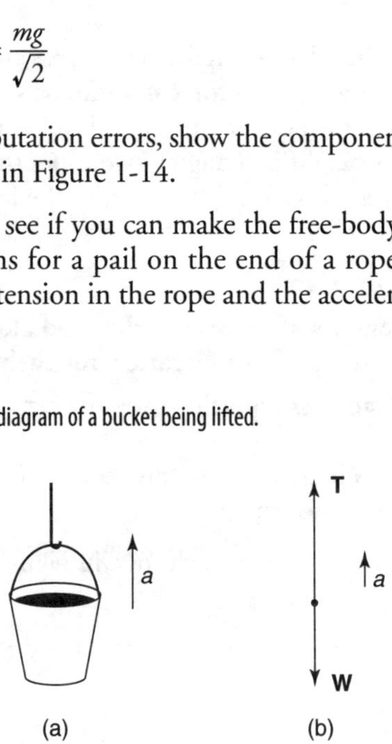

The two forces acting on the pail are the tension of the rope (**T**) and weight (**W** = *mg*). By Newton's second law:

Solution: $F_{net} = T - mg = ma$

Example 3: Next, try to set up the equations for a two-body system of unequal masses attached by a rope over a frictionless pulley (see Figure 1-16). A diagram must be made for each of the two objects of mass (m_1) and (m_2). Find the acceleration of the system and the tension in the rope.

Figure 1-16 Force diagram of two masses hung over a pulley.

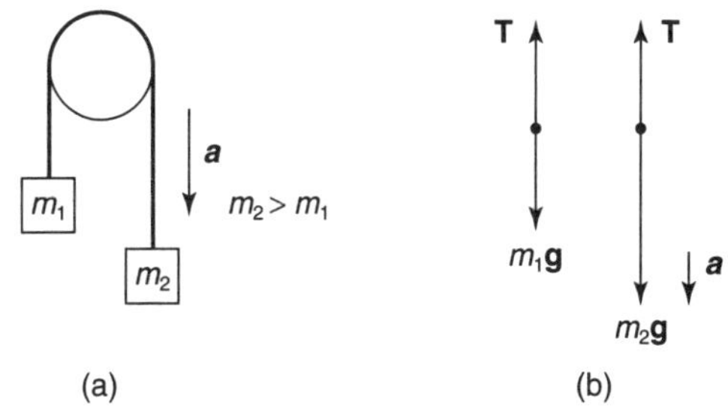

For m_1: $F_{net} = T - m_1 g = m_1 a$, as magnitudes

For m_2: $T - m_2 g = - m_2 a$, as magnitudes

The second equation has a negative acceleration because m_2 is descending. Because the objects are connected by a rope that does not expand, the tensions and accelerations are the same for each mass. By algebraic manipulation, the equations may be simultaneously solved for the following results:

Solution: $a = \frac{m_2 - m_1}{m_2 + m_1} g$ and $T = \frac{2m_1 m_2}{m_1 + m_2} g$

Example 4: In Figure 1-17, one object sits on a frictionless surface, and the other object hangs off the edge of the table over a pulley. Make the free-body diagram and write the force equations to find the acceleration and tension.

Figure 1-17 Same as Figure 1-16 with one mass on a frictionless table.

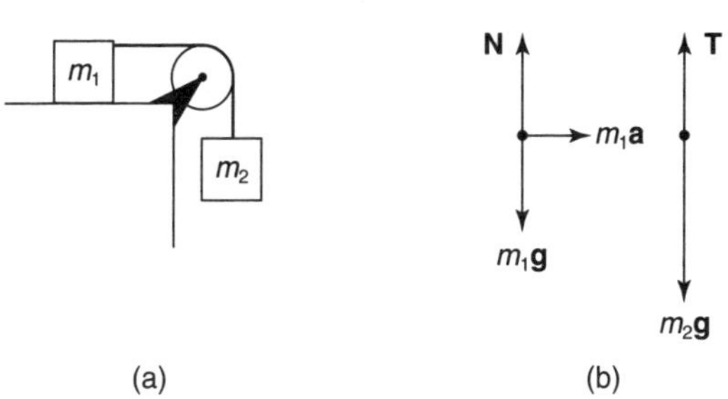

In the vertical direction, the forces on m_1 are the weight downward and the normal force (**N**) upward due to the surface, which is equal and opposite to the weight. Because there is no acceleration of m_1 in this direction, the net vertical force is zero. A horizontal force of the tension in the rope accelerates the mass to the right. Write the force equations separately for x and

y directions for m_1. The forces on the second mass are the same as those in the last example. For m_1 in the x direction: $T = m_1 a$; in the y direction: $N - m_1 g = 0$. For m_2: $T - m_2 g = -m_2 a$. Combining the two equations gives the relationships.

Solution: $a = \frac{m_2}{m_1 + m_2} g$ and $T = \frac{m_1 m_2}{m_1 + m_2} g$

Even more complicated problems can be separated into manageable parts to allow solution by using these problem-solving methods.

Example 5: Consider a mass (m_1) on an inclined plane attached to a mass (m_2) over a pulley as in Figure 1-18(a). Both the plane and pulley are frictionless. Set up the problem to find the acceleration.

Figure 1-18 Now the table in Figure 1-17 is tilted at an angle.

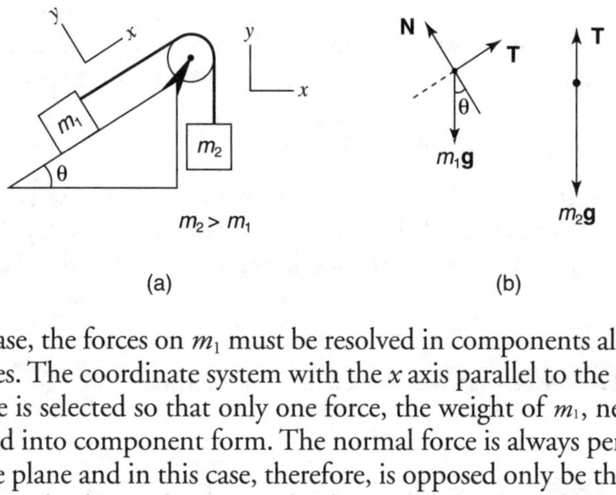

In this case, the forces on m_1 must be resolved in components along the x and y axes. The coordinate system with the x axis parallel to the surface of the plane is selected so that only one force, the weight of m_1, needs to be converted into component form. The normal force is always perpendicular to the plane and in this case, therefore, is opposed only be the component of weight that is also perpendicular to the plane's surface. Note that the angle between the y axis and the weight ($m_1 \mathbf{g}$) is the angle of inclination of the plane, which can be proven by geometry. The coordinate system for m_2 has the same orientation as in the previous example. Assume $m_2 > m_1$. Then, for m_1 in its y direction: $N - m_1 g \cos\theta = 0$; in its x direction: $T - m_1 g \sin\theta = m_1 a$. For m_2: $T - m_2 g = -m_2 a$. The acceleration is then

Solution: $a = \frac{(m_2 - m_1 \sin\theta) g}{m_1 + m_2}$

and tension is $T = \frac{(1 + \sin\theta) m_1 m_2 g}{m_1 + m_2}$

To analyze a physical situation by the use of free-body diagrams, use the following steps:

1. **Make a free-body diagram for each object.** If one object is sitting on a surface, be sure to include the normal force.
2. **Resolve the forces that are not directed along the x and y axes into components along a preferred coordinate system.** For inclined planes, use a coordinate axis with the x axis parallel to the surface of the plane. Put the components on a separate diagram; that is, do not put the force and its components on the same diagram because this combination might complicate the following steps.
3. **Write out the force equation for each mass along each axis, noting the correct sign for the acceleration of the body.**
4. **Solve the equations simultaneously to find the desired value(s).**

Friction

Friction is the force opposing the motion of one body sliding or rolling over the surface of second object. Several aspects of friction are important at low velocities:

- The direction of the force of friction is opposite the direction of motion.
- The frictional force is proportional to the perpendicular (normal) force between the two surfaces in contact.
- The frictional force is nearly independent of the area of contact between the two objects.
- The magnitude of the frictional force depends on the materials composing the two objects in contact.

Static friction is the force of friction when there is no relative motion between two objects in contact, such as a block sitting on an inclined plane. The magnitude of the frictional force is $F_s \leq \mu_s N$, where N is the magnitude of the normal force, and the coefficient of friction (μ_s) is the dimensionless proportionality constant. The **coefficient of friction** may be approximately .2 for normal lubricated surfaces and close to 1 for glass sliding on glass. This equation sets the upper limit for the static frictional force. If a greater external force is applied, the situation will no longer be static, and the object will begin to move.

Kinetic friction is the force of friction when there is relative motion between two objects in contact. The magnitude of the friction force in this case is $F_k \leq \mu_k N$, where N is the magnitude of the normal force and μ_k is the coefficient of kinetic friction. Note that μ_k is not strictly a constant, but this empirical rule is a good approximation for finding frictional forces. Values given for the coefficients of static and kinetic friction do vary with speed and surface conditions so that it is not necessarily true that static friction exceeds sliding friction.

The following problem highlights the differences between static and kinetic friction.

Example 6: A block sits on an inclined plane. What is the maximum angle that which the block remains at rest? First, draw the free-body diagrams and then write out the force equation for each direction of the coordinate system (see Figure 1-19).

Figure 1-19 A block on an inclined plane, with friction.

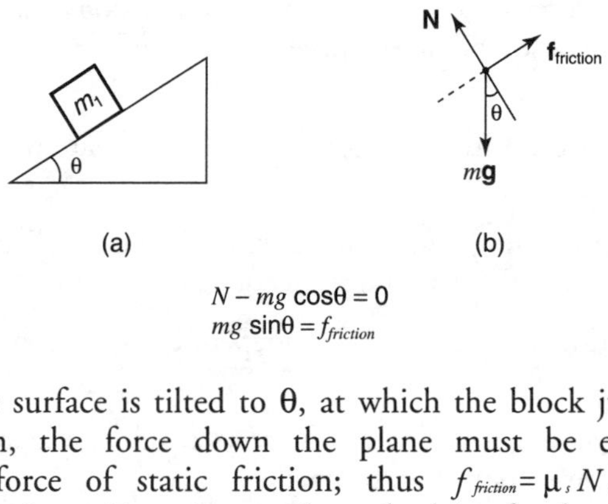

$$N - mg\cos\theta = 0$$
$$mg\sin\theta = f_{friction}$$

Suppose the surface is tilted to θ, at which the block just begins to move. Then, the force down the plane must be equal to the maximum force of static friction; thus $f_{friction} = \mu_s N$. Therefore, $f_{friction} = mg \sin\theta = \mu_s N = \mu_s(mg\cos\theta)$ and solving for the coefficient of friction:

Solution: $\mu_s = \tan\theta$

At a greater angle of tilt, the object accelerates down the surface, and the force of friction is $f_k = \mu_k N$.

Example 7: If the surfaces in Figures 1-17 and 1-18 were *not* frictionless, the frictional force parallel to the surface and opposite the direction of motion must be included in the analysis. Pulling a block along a horizontal surface at a constant speed (zero acceleration) is an example of a problem involving friction, and such a block is analyzed in Figure 1-20.

Figure 1-20 Pulling a block on a plane with friction.

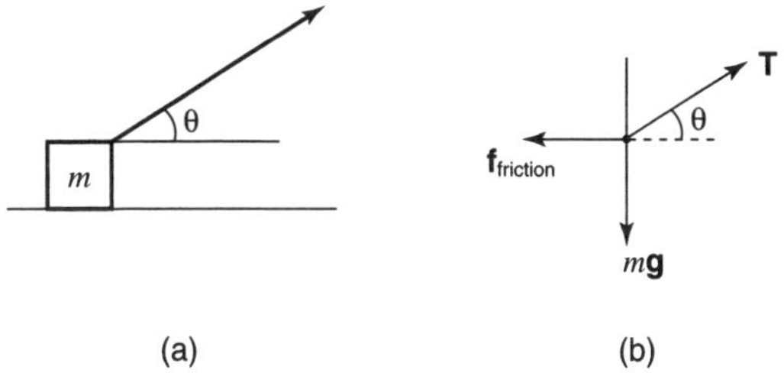

In the x direction, $T\cos\theta - f = 0$, where f is the friction force and T is the tension in the rope. In the y direction, $N + T\sin\theta - mg = 0$; also, $f = \mu_k N$.

Solve the y direction equation for N, substitute the expression into the friction force equation, and then substitute friction into the first equation to obtain the following:

$$T\cos\theta - \mu_k(mg - T\sin\theta) = 0$$

Solution: $T = \dfrac{\mu_k mg}{\cos\theta + \mu\sin\theta}$, solving for T.

Centripetal force

When an object rotates in a circle, a force must be directed to the centre of the circle to maintain the motion; otherwise, the object will take off tangent to the path. This constraining force is called a **centripetal force,** meaning centre-seeking. In the example of a mass rotating in a horizontal circle at the end of the string, the centripetal force is provided by the tension in the string. In the case of orbiting satellites, gravity provides the centre-seeking force. From the definition of force $\mathbf{F} = m\mathbf{a}$ and the expressions for circular accelerations, the following equations are obtained:

$$F_c = m\frac{v^2}{R} \text{ or } F_c = m\frac{2\pi v}{T} \text{ or } F_c = \frac{m4\pi^2 - R}{T^2}$$

If an object moves in a circle, the net force is a centripetal force. One such example is the conic pendulum, a mass on the end of a string that rotates in a horizontal circle (see Figure 1-21).

Figure 1-21 A conic pendulum.

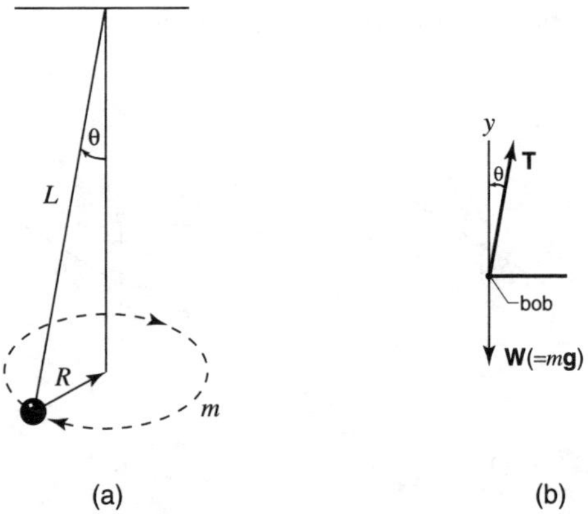

In the y direction, $T \cos\theta - mg = 0$. In the x direction (or the radial direction), $T \sin\theta = mv^2/R$, where R is the radius of circular path. Dividing the second equation by the first equation and solving for v yields the following:

$$v = \sqrt{\frac{gR \sin\theta}{\cos\theta}} = \sqrt{gR \tan\theta}$$

Universal gravitation

Newton's law of universal gravitation states that every mass in the universe attracts every other mass with a gravitational force that is directly proportional to the product of their masses (m_1, m_2) and inversely proportional to the square of distance *(r)* between them. In mathematical form, $F = (Gm_1 m_2)/r^2$, where G is the **universal gravitational constant.** In the metric system, the accepted value of G is $G = 6.673 \times 10^{-11} (Nm^2/kg^2)$.

Kepler found three empirical laws regarding the motion of satellites that Newton later showed from his law of universal gravitation. These are **Kepler's laws of planetary motion:**

- **The law of orbits:** All planets move in elliptical orbits with the sun at one focus.
- **The law of areas:** A line joining a planet and the sun sweeps out equal areas in equal time.
- **The law of periods:** The square of the period *(T)* of any planet is proportional to the cube of the semi-major axis (r) of its orbit, or $T^2 = (4\pi^2/GM)\, r^3$, where M is the mass of the planet.

Momentum and impulse

According to Newton's second law, a mass experiencing a net average force (**F**) for a time interval Δt will undergo an average acceleration (**F**=m**a**). The product of the average force acting on the body and time of contact is defined as **impulse.** Because acceleration is change in velocity, the relationships between these variables are expressed as the $\text{Impulse} = \mathbf{F}(\Delta t) = m\mathbf{v}_f - m\mathbf{v}_i$, where $\mathbf{v}_i$ is the initial velocity and $\mathbf{v}_f$ is the velocity after the force is no longer in contact with the body. Impulse is measured in units of newton-seconds, or more simply, N-s.

When applying the impulse equation, be sure to calculate the vector change in velocity—for example, consider a mass of 10 kg acted on by a force that changes its velocity from –8 m/s to 3 m/s . This force imparted an impulse of (10 kg)[3 – (–8) m/s] = (10 kg)(11 m/s) = 110 N-s.

The right side of the impulse equation is the change in the **linear momentum** of the object. The definition of linear momentum is $\mathbf{p} = m\mathbf{v}$. Linear momentum is measured in units of kilogram metres/second or, in abbreviated form, kg m/s. Newton originally stated his second law by saying that the rate of change of momentum with time is proportional to the impressed force and is in the same direction; thus, $\mathbf{F} = \Delta(m\mathbf{v})/\Delta t$ or $\mathbf{F} = \Delta\mathbf{p}/\Delta t$.

Conservation of momentum

An extremely important fundamental principle in physics is the law of **conservation of momentum**. The law states that if there is no external force acting on a system, the total momentum remains a constant, which provides a powerful way to analyze interactions between systems of objects. For example, if a rolling ball on a frictionless surface collides with another ball, the total momentum before and after the collision is the same. An interaction, therefore, can be examined without knowing the forces involved and the length of interaction time, which might be difficult to measure.

First, consider a head-on collision (so it is not necessary to utilize two-dimensional vectors, that is, consider only straight line motion). Imagine a mass (m_1) with velocity ($\mathbf{v}_1$) hitting a mass (m_2) that is initially at rest. The momentums are the following: before the collision, $m_1\mathbf{v}_1$, and after the collision, $m_1\mathbf{v}_1' + m_2\mathbf{v}_2'$, where the primes indicate velocities after the interaction. From the law of conservation of momentum, the two expressions may be set equal to each other. Consider the special case where the two masses are equal on frictionless surface and stick together after the collision (have the same primed velocity). Then, total momentum before the collision equals total momentum after the collision, $m\mathbf{v}_1 = m\mathbf{v}' + m\mathbf{v}'$; therefore, $\mathbf{v}' = (1/2)(\mathbf{v}_1)$, or the final velocity is one-half the original velocity because the effective mass has doubled.

Another way to state the law of conservation of momentum is that the change in momentum of the two objects must be equal and opposite. For example, two ice skaters are at rest in the centre of frictionless ice (possible at least in the imagination). Let one have a relatively small mass *(m)* and the other a larger mass *(M)*. Because they begin at rest, the initial momentum is zero. They then push apart in opposite directions. The total momentum must remain zero.

According to the law of conservation of momentum, $\Delta \mathbf{p}_m = -\Delta \mathbf{p}_M$ or $m\mathbf{v}' - 0 = -(M\mathbf{V}' - 0)$; therefore, if the large mass *(M)* is three times the smaller mass *(m)*, $\mathbf{v}' = -3\mathbf{V}'$, where $\mathbf{v}'$ is the velocity of the small mass after the collision and $\mathbf{V}'$ is the velocity of the large mass after the collision. The negative sign indicates velocities in opposite directions.

This same analysis holds for a person standing on frictionless ice who throws an object; it even holds for a rocket going to the moon. The ice skater throwing a glove attains equal momentum in the direction opposite to that of the thrown object. This basic principle is the same for a rocket accelerating in space. Spacecrafts utilize the law of conservation of momentum in getting an additional push from discharged rocket stages as well as from fuel. In particular, the Apollo space capsule returning from the moon was only a small per cent of the total mass initially sent upward from the launch pad; therefore, acceleration of a rocket can be caused by either a change in velocity, by a change in its mass, or by changes in both velocity and mass (see Figure 1-22). Thus, the expression of Newton's second law of motion, stated in terms of the change of momentum, is broader than the expression given only in terms of mass and acceleration.

Figure 1-22 A rocket ship gains momentum.

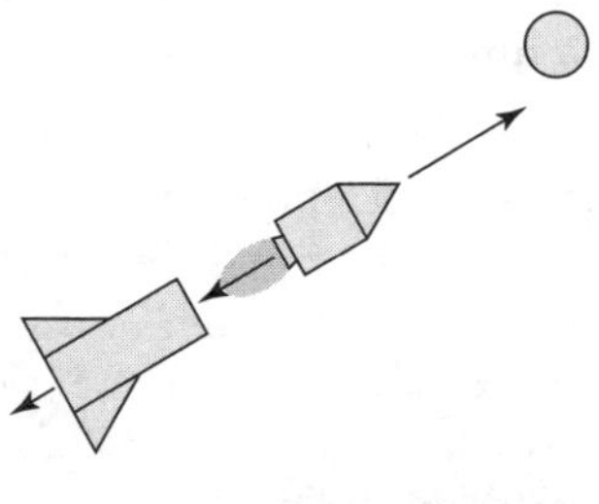

If two objects strike with a glancing blow, the motion will be two-dimensional. For example, one ball (m_1) with an initial velocity hits a second ball (m_2), which is initially at rest. Figures 1-23 and 1-24 depict this situation with the first ball initially moving up from the bottom of the page. For the sake of simplicity, allow the two masses to be equal.

The momentum vectors can be added to show the law of conservation of momentum. The vector addition in Figure 1-24(a) shows that the total of the two momentum vectors, $\mathbf{p}_1'$ and $\mathbf{p}_2'$, after the collision are equal to the total momentum before the collision. (Because only m_1 was moving, there was only one initial momentum vector, $\mathbf{p}_1$.) Figure 1-24(b) also shows the alternate method of using the law of conservation of momentum, that the change (difference) in momentum of m_1 is equal and opposite to that of m_2.

Figure 1-23 Two balls striking a glancing blow.

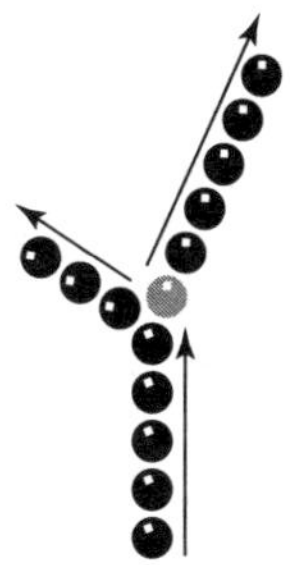

Figure 1-24 (a) Total momentum is conserved. (b) Equal and opposite momentum changes of the two balls.

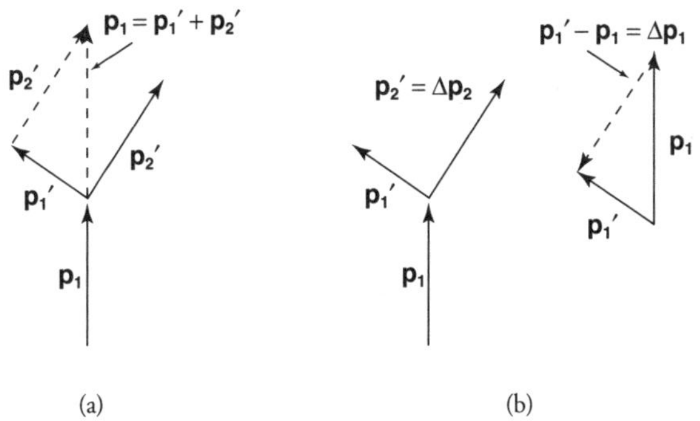

If the two masses are not equal, the velocity vectors must be adjusted so that the vectors represent momentum. For example, if one mass is three times the other, the velocity vectors of the larger mass must be lengthened by a factor of three before using the law of conservation of momentum.

Work and Energy

The concepts of work and energy are closely tied to the concept of force because an applied force can do work on an object and cause a change in energy. **Energy** is defined as the ability to do work.

Work

The concept of work in physics is much more narrowly defined than the common use of the word. **Work** is done on an object when an applied force moves it through a distance. In our everyday language, work is related to expenditure of muscular effort, but this is *not* the case in the language of physics. A person that holds a heavy object does no physical work because the force is not moving the object through a distance. Work, according to the physics definition, is being accomplished while the heavy object is being lifted but not while the object is stationary. Another example of the absence of work is a mass on the end of a string rotating in a horizontal circle on a frictionless surface. The centripetal force is directed toward the centre of the circle and, therefore, is not moving the object through a distance; that is, the force is not in the direction of motion of the object. (However, work was done to set the mass in motion.) Mathematically, work is $W = \mathbf{F} \cdot \mathbf{x}$, where **F** is the applied force and **x** is the distance moved, that is, displacement. Work is a scalar. The SI unit for work is the joule (J), which is newton-metre or kg m/s^2.

If work is done by a varying force, the above equation cannot be used. Figure 1-25 shows the force-versus-displacement graph for an object that has three different successive forces acting on it. The force is increasing in segment I, is constant in segment II, and is decreasing in segment III. The work performed on the object by each force is the area between the curve and the *x* axis. The total work done is the total area between the curve and the *x* axis. For example, in this case, the work done by the three successive forces is shown in Figure 1-25.

Figure 1-25 Acting force changing with position.

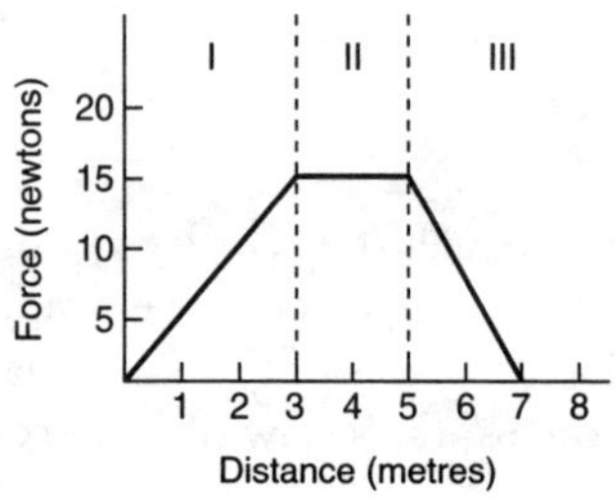

In this example, the total work accomplished is (1/2)(15)(3) + (15)(2) + (1/2)(15)(2) = 22.5 + 30 + 15; work = 67.5 J. For a gradually changing force, the work is expressed in integral form, $W = \int \mathbf{F} \cdot d\mathbf{x}$.

Kinetic energy

Kinetic energy is the energy of an object in motion. The expression for kinetic energy can be derived from the definition for work and from kinematic relationships. Consider a force applied parallel to the surface that moves an object with constant acceleration.

From the definition of work, from Newton's second law of motion, and from kinematics, $W = Fx = max$ and $v_f^2 = v_o^2 + 2ax$, or $a = (v_f^2 - v_o^2)/2x$. Substitute the last expression for acceleration into the expression for work to obtain $W = m(v_f^2 - v_o^2)/2$ or $W = (1/2)mv_f^2 - (1/2)mv_o^2)$. The right side of the last equation yields the definition for kinetic energy: $K.E. = (1/2)mv^2$. Kinetic energy is a scalar quantity with the same units as work, joules (J). For example, a 2 kg mass moving with a speed of 3 m/s has a kinetic energy of 9 J.

The above derivation shows that the net work is equal to the change in kinetic energy. This relationship is called the **work-energy theorem:** $W_{net} = K.E._f - K.E._o$, where $K.E._f$ is the final kinetic energy and $K.E._o$ is the original kinetic energy.

Potential energy

Potential energy, also referred to as stored energy, is the ability of a system to do work due to its position or internal structure. Examples are energy stored in a pile driver at the top of its path or energy stored in a coiled spring. Potential energy is measured in units of joules.

Gravitational potential energy is energy of position. First, consider gravitational potential energy near the surface of the earth where the acceleration due to gravity *(g)* is approximately constant. In this case, an object's gravitational potential energy with respect to some reference level is $P.E. = mgh$, where h is the vertical distance above the reference level. To lift an object slowly, a force equal to its weight *(mg)* is applied through a height *(h)*. The work accomplished is equal to the change in potential energy: $W = P.E._f - P.E._o = mgh_f - mgh_o$, where the subscripts *(f* and *o)* refer to the final and original heights of the body.

Launching a rocket into space requires work to separate the mass of the earth and the rocket to overcome the gravitational force. For large distances from the centre of the earth, the above equation is inadequate because

g is not constant. The **general form of gravitational potential energy** is $P.E. = -GMm/r$, where M and m refer to the masses of the two bodies being separated and r is the distance between the centres of the masses. The negative sign is a result of selecting the zero reference at r equal to infinity, that is, at very large **separation.**

Elastic potential energy is energy stored in a spring. The magnitude of the force required to stretch a spring is given by $F = -kx$, where x is the distance of stretch (or compression) of a spring from the unstressed position, and k is the **spring constant.** The spring constant is a measure of the stiffness of the spring, with stiffer springs having larger k values. The potential energy stored in a spring is given by $P.E. = (1/2)\, kx^2$.

Change in potential energy is equal to work. The gravitational force and the force to stretch a spring are varying forces; therefore, the potential energy equations given above for these two cases can also be derived from the integral form of work, $\Delta P.E. = W = \int \mathbf{F} \cdot d\mathbf{x}$.

Power

Power is the rate of doing work, average $P = W/t$, where t is the time interval during which work (W) is accomplished. Another form of power is found from $W = F\,\Delta x$ and substitution of average velocity of the object during time t for $\Delta x / t$: average $P = F\,\Delta x / \Delta t = F(\text{average } v)$.

The conservation of energy

The principle of **conservation of energy** is one of the most far-reaching general laws of physics. It states that energy is neither created nor destroyed but can only be transformed from one form to another in an isolated system.

Because the total energy of the system always remains constant, the law of conservation of energy is a useful tool for analyzing a physical situation where energy is changing form. Imagine a swinging pendulum with negligible frictional forces. At the top of its rise, all the energy is gravitational potential energy due to height above the stationary position. At the bottom of the swing, all the energy has been transformed into kinetic energy of motion. The total energy is the sum of the kinetic and potential energies. It maintains the same value throughout the back and forth motion of a swing (see Figure 1-26).

Figure 1-26 A pendulum obeys the law of conservation of energy.

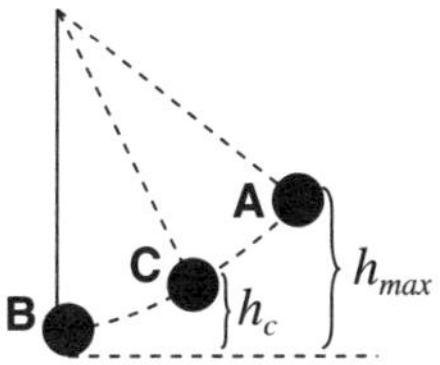

At point A: *total energy* = $P.E._{max} = mgh_{max}$

At point B: *total energy* = $K.E._{max} = \frac{1}{2}mv^2_{max}$

At point C: *total energy* = $mgh_c + \frac{1}{2}mv_c^2$

At point C, the potential energy is dependent upon the height, and the rest of the total energy is kinetic energy.

Elastic and inelastic collisions

Although total energy is conserved, kinetic energy need not be conserved. A collision between two objects with conservation of kinetic energy is called an **elastic collision.** Colliding objects interacting with losses of kinetic energy due to frictional losses or deformation of an object are called **inelastic collisions.** In the macroscopic world, most collisions are inelastic; however, losses of kinetic energy are negligible in the nearly elastic collisions between atomic particles and subatomic particles. For these cases, the law of conservation of momentum and the conservation of kinetic energy yield useful equations.

Consider a simple head-on elastic collision where one mass (m_1) with a given velocity (v_1) hits a second mass (m_2) that is initially at rest. Apply the laws of conservation of momentum and conservation of kinetic energy to get $m_1v_1 = m_1v_1' + m_2v_2'$ and $(1/2)m_1v_1^2 = (1/2)m_1v_1'^2 + (1/2)m_2v_2'^2$, where the primes refer to velocities after the collision. Solving the equations gives the velocities of the two masses after the interaction:

$$v_1' = \frac{m_1 - m_2}{m_1 + m_2}v_1 \qquad v_2' = \frac{2m_1}{m_1 + m_2}v_1$$

Three special cases are instructive:

- For equal masses where $m_1 = m_2$, note that v_1' becomes zero and v_2' equals v_1; thus, for equal masses, the objects simply exchange velocities, as is sometimes observed with pool balls. (Pool balls have rotational energy and somewhat inelastic collisions, so their behaviour only approximates the example.)

- If m_2 is massive, the numerator and denominator are nearly the same in the equation for v_1'. Then, v_1' is approximately equal to v_1 but in the opposite direction. The denominator of the expression for v_2' will be so large so that the velocity of the second mass after the collision will be small. In other words, the incoming mass (m_1) will bounce back off the second mass with nearly the initial speed, and the hit mass (m_2) will move slowly after the collision.
- If m_1 is massive, then v_1' is approximately equal to v_1, and v_2' is nearly twice v_1; or the incoming massive particle continues at nearly the same velocity, and the hit mass moves ahead at nearly twice the initial velocity of the first mass after the collision.

Centre of mass

The concept of the **centre of mass** (CM) is useful to analyze the motion of a system of particles. The system of particles acts as if all of its mass is concentrated at the CM. In the absence of an external force, if the CM of the system is at rest, then it will remain at rest, and if it is initially in motion, it will maintain that motion. In other words, the CM moves in accordance with Newton's second law. The *x* and *y* coordinates of the centre of mass are

$$x_{CM} = \frac{\sum m_i x_i}{\sum m_i} \text{ and } y_{CM} = \frac{\sum m_i y_i}{\sum m_i}$$

Consider the previous example of a head-on collision of two equal masses that stick together after the collision. The CM is initially in motion at a constant velocity and maintains the same velocity after the collision. As the first mass rolls in toward the second mass, the CM is always halfway between the two masses. Before the collision, the CM covers one-half the distance of the incoming object in the same time, and therefore, the velocity of the CM is one-half the initial velocity of the incoming mass. For the instant that the two masses interact, the CM is right between the two objects. After the collision, the masses stick together and have one-half the initial velocity because the effective mass has doubled. The CM continues midway between the masses. It maintains the same velocity of (1/2) v_o after the collision. In Figure 1-27, the moving white ball impacts the stationary black ball. The numbered and circled positions of the CM correspond to the numbered positions of the balls.

Figure 1-27 Inelastic collision of two balls.

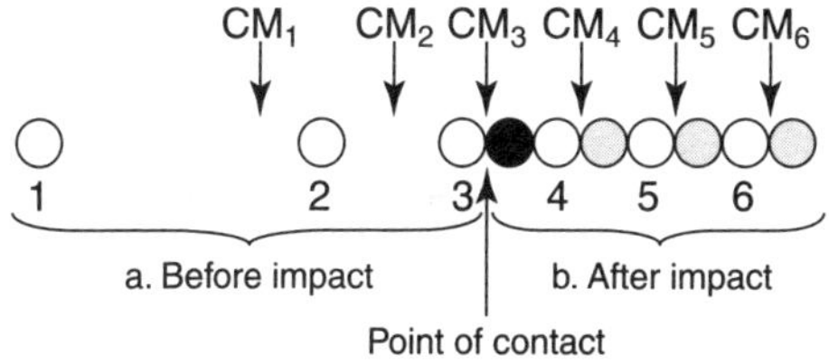

Rotational Motion of a Rigid Body

Rotational motion is more complicated than linear motion, and only the motion of rigid bodies will be considered here. A **rigid body** is an object with a mass that holds a rigid shape, such as a phonograph turntable, in contrast to the sun, which is a ball of gas. Many of the equations for the mechanics of rotating objects are similar to the motion equations for linear motion.

Angular velocity and angular acceleration

The **angular displacement** of a rotating wheel is the angle between the radius at the beginning and the end of a given time interval. The SI units are radians. The average **angular velocity** (ω, Greek letter omega), measured in radians per second, is

$$\omega = \frac{\text{angular displacement}}{\text{elapsed time}} = \frac{\theta}{t}$$

The **angular acceleration** (α, Greek letter alpha) has the same form as the linear quantity

$$\alpha = \frac{\text{change in angular velocity}}{\text{elapsed time}} = \frac{\omega_f - \omega_o}{t}$$

and is measured in radians/second/second or rad/s^2.

The kinematics equations for rotational motion at constant angular acceleration are

$\omega_f = \omega_o + \alpha t$	Angular velocity as a function of time
$\theta = \frac{1}{2}(\omega_o + \omega_f)\,t$	Angular displacement as a function of velocity and time
$\theta = \omega_o t + \frac{1}{2}\alpha t^2$	Angular displacement as a function of time
$\omega_f^2 = \omega_o^2 + 2\alpha\theta$	Angular velocity as a function of displacement

Consider a wheel rolling without slipping in a straight line. The forward displacement of the wheel is equal to the linear displacement of a point fixed on the rim. As can be shown in Figure 1-28, $d = S = r\theta$.

In this case, the average forward speed of the wheel is $v = d/t = (r\theta)/t = r\omega$, where *r* is the distance from the centre of rotation to the point of the calculated velocity. The direction of the velocity is tangent to the path of the point of rotation.

Figure 1-28 A wheel rolling without slipping.

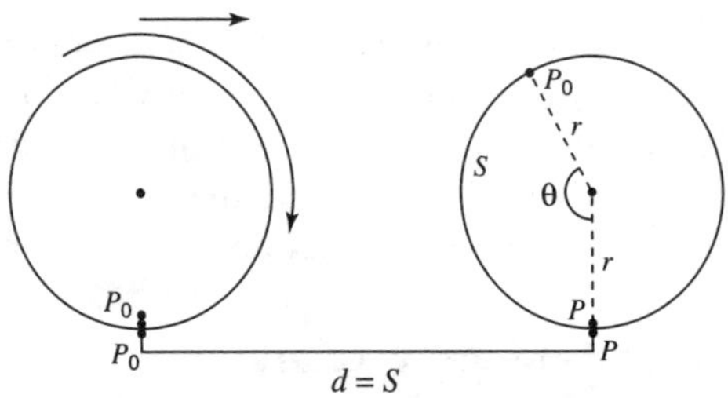

The average forward acceleration of the wheel is $a_T = r(\omega_f - \omega_0)/t = r\alpha$. This component of the acceleration is tangential to the point of rotation and represents the changing speed of the object. The direction is the same as the velocity vector.

The radial component of the linear acceleration is $a_r = v^2/r = \omega^2 r$.

Torque

It is easier to open a door by pushing on the edge farthest from the hinges than by pushing in the middle. It is intuitive that the magnitude of the force applied and the distance from the point of application to the hinge affect the tendency of the door to rotate. This physical quantity, **torque,** is $\mathbf{t} = \mathbf{r} \times \mathbf{F} \sin\theta$, where *F* is the force applied, *r* is the distance from the point of application to the centre of the rotation, and θ is the angle from *r* to *F*.

Moment of inertia

Substitute Newton's second law into the definition for torque with θ of 90 degrees (a right angle between *F* and *r*) and use the relationship between linear acceleration and tangential angular acceleration to obtain $t = rF = rma = mr^2(a/r) = mr^2\alpha$. The quantity mr^2 is defined as **moment of inertia** of a point mass about the centre of rotation.

Imagine two objects of the same mass with different distribution of that mass. The first object might be a heavy ring supported by struts on an axle like a flywheel. The second object could have its mass close to the central axis. Even though the masses of the two objects are equal, it is intuitive that the flywheel will be more difficult to push to a high number of revolutions per second because not only the amount of mass but also the distribution of the mass affects the ease in initiating rotation for a rigid body. The general definition of moment of inertia, also called **rotational inertia,** for a rigid body is $I = \sum m_i r_i^2$ and is measured in SI units of kilogram-metres^2.

The moments of inertia for different regular shapes are shown in Figure 1-29.

Figure 1-29 Moments of inertia for various regular shapes.

A hoop or cylindrical shell about the axis:	$I = mr^2$	
Disc or solid cylinder about an axis through the centre:	$I = \frac{1}{2}mr^2$	
A solid sphere about any diameter:	$I = \frac{2}{5}mr^2$	
A thin rod about an axis perpendicular through the centre:	$I = \frac{1}{12}ml^2$	

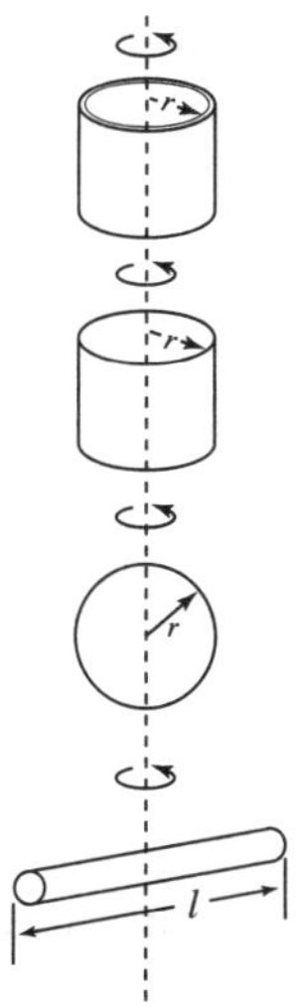

Mechanics problems frequently include both linear and rotation motions.

Example 8: Consider Figure 1-30, where a mass is hanging from a rope wrapped around a pulley. The falling mass *(m)* causes the pulley to rotate, and it is no longer necessary to require the pulley to be massless. Assign mass (*M*) to the pulley and treat it as a rotating disc with radius *(R)*. What is the acceleration of the falling mass, and what is the tension of the rope?

Figure 1-30 A hanging mass spins a pulley.

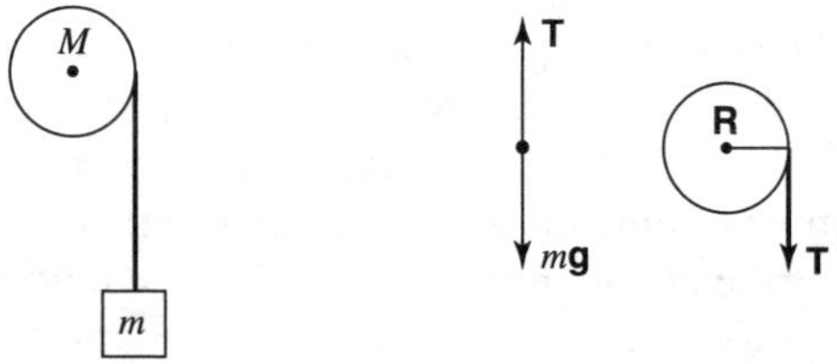

(a) (b)

The force equation for the falling mass is $T - mg = -ma$. The tension of the rope is the applied force to the edge of the pulley that is causing it to rotate. Thus, $t = I\alpha$, or $TR = (1/2)MR^2(a/R)$, which reduces to $T = (1/2)Ma$, where angular acceleration has been replaced by a/R because the cord does not slip and the linear acceleration of the block is equal to the linear acceleration of the rim of the disk. Combining the first and last equation in this example leads to

Solution: $a = g\dfrac{2m}{M+2m}$ and $T = mg\dfrac{M}{M+2m}$

Angular momentum

Angular momentum is rotational momentum that is conserved in the same way that linear momentum is conserved. For a rigid body, the angular momentum *(L)* is the product of the moment of inertia and the angular velocity: $L = I\omega$. For a point of mass, angular momentum can be expressed as the product of linear momentum and the radius (r): $L = mvr$. L is measured in units of kilograms-metres^2 per second or more commonly joule-seconds. The **law of conservation of angular momentum** can be stated that the angular momentum of a system of objects is conserved if there is not external net torque acting on the system.

Analogous to Newton's law ($\mathbf{F} = \Delta(m\mathbf{v})/\Delta t$) there is a rotational counterpart for rotational motion: $t = \Delta L/\Delta t$, or torque is the rate of change of angular momentum.

Consider the example of a child who runs tangential to the edge of a playground merry-go-round with a velocity v_o and jumps on while the merry-go-round is at rest. The only external forces are that of gravity and the contact forces provided by the support bearings, neither of which causes a torque because they are not applied to cause a horizontal rotation. Treat the child's mass as a point of mass and the merry-go-round as a disc with a radius R and mass M. From the conservation law, the total angular momentum of the child before the interaction is equal to the total angular momentum of the child and merry-go-round after the collision: $mrv_o = mrv' + I\omega$, where r is the radial distance from the centre of the merry-go-round to the place where the child hits. If the child jumps on the edge, $(r = R)$ and the angular velocity for the child after the collision can be substituted for the linear velocity, $mRv_o = mR(R\omega) + (1/2)MR^2$. If the values for the masses and the initial velocity of the child are given, the final velocity of the child and merry-go-round can be calculated.

A single object may have a change in angular velocity due to the conservation of angular momentum if the distribution of the mass of the rigid body is altered. For example, when a figure skater pulls in her extended arms, her moment of inertia will decrease, causing an increase in angular velocity. According to the conservation of angular momentum, $I_o(\omega_o) = I_f(\omega_f)$, where I_o is the moment of inertia of the skater with arms extended, I_f is her moment of inertia with her arms close to her body, ω_o is her original angular velocity, and ω_f is her final angular velocity.

Rotational kinetic energy, work, and power. Kinetic energy, work, and power are defined in rotational terms as $K.E. = (1/2)I\omega^2$, $W = t\theta$, $P = t\omega$.

Comparison of dynamics equation for linear and rotational motion. The dynamic relations are given to compare the equation for linear and rotational motion (see Table 1-2).

Table 1-2 Comparison of Dynamics Equations for Linear and Rotational Motion

	Linear Motion	*Rotational Motion*
Newton's second law	$F = ma$	$t = I\alpha$
Momentum	$p = mv$	$L = I\omega$
Work	$W = F\Delta x$ or $W = \int \mathbf{F} \cdot \mathbf{dx}$	$W = t\Delta\theta$ or $W = \int \mathbf{t} \cdot \mathbf{d}\theta$
Kinetic energy	$K.E. = \frac{1}{2}mv^2$	$K.E. = \frac{1}{2}I\omega^2$
Power	$P = Fv$	$P = t\omega$

Elasticity and Simple Harmonic Motion

A rigid body is an idealization because even the strongest material deforms slightly when a force is applied. **Elasticity** is the field of physics that studies the relationships between solid body deformations and the forces that cause them.

Elastic modules

In general, an **elastic modulus** is the ratio of stress to strain. Young's modulus, the bulk modulus, and the shear modulus describe the response of an object when subjected to tensile, compressional, and shear stresses, respectively.

When an object such as a wire or a rod is subjected to a tension, the object's length increases. **Young's modulus** is defined as the ratio of tensile stress and tensile strain. **Tensile stress** is a measure of the deformation that causes stress. Its definition is the ratio of tensile force *(F)* and the cross-sectional area normal to the direction of the force *(A)*. Units of stress are newtons per square metre (N/m^2). **Tensile strain** is defined as the ratio of the change in length $(l - l_o)$ to the original length (l_o). Strain is a number without units; therefore, the expression for Young's modulus is

$$Y = \frac{F/A}{(l - l_o)/l_o}$$

If an object of cubic shape has a force applied pushing each face inward, a compressional stress occurs. **Pressure** is defined as force per area $P = F/A$. The SI unit of pressure is the pascal, which is equal to 1 newton/metre^2 or N/m^2. Under uniform pressure, the object will contract, and its fractional change in volume *(V)* is the **compressional strain.** The corresponding elastic modulus is called the **bulk modulus** and is given by $B = -P/(\Delta V/V_o)$. The negative sign ensures that *B* is always a positive number because an increase in pressure causes a decrease in volume.

Applying a force on the top of an object that is parallel to the surface on which it rests causes a deformation. For example, push the top of a book resting on a tabletop so that the force is parallel to surface. The cross-section shape will change from a rectangle to a parallelogram due to the **shear stress** (see Figure 1-31). Shear stress is defined as the ratio of the tangential force to the area *(A)* of the face being stressed. **Shear strain** is the ratio of the horizontal distance the sheared face moves (Δx) and the height of the object *(h)*, which leads to the **shear modulus:**

$$S = \frac{F/A}{\Delta x/h}$$

Figure 1-31 Shear stress deforms a book.

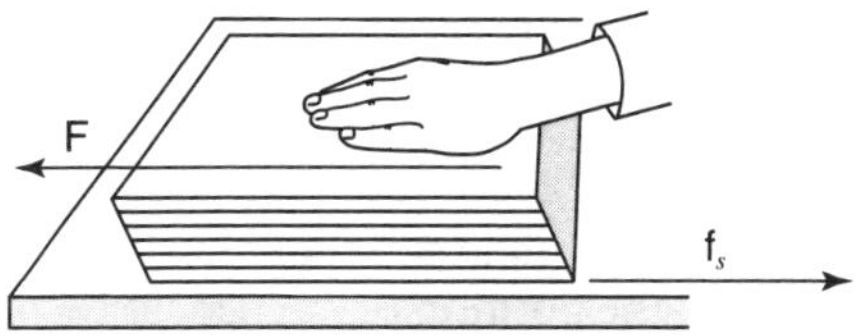

Hooke's Law

The direct relationship between an applied force and the change in length of a spring, called **Hooke's law,** is $F = -kx$, where x is the stretch in the spring and k is defined as the **spring constant.** Units for k are newtons per metre. When a mass is hung on the end of the spring, at equilibrium the downward gravitational force on the mass must be balanced by an upward force due to the spring. This force is called the **restoring force.** The negative sign indicates that the direction of the restoring force due to the spring is in the opposite direction from the stretch, or displacement, of the spring.

Simple harmonic motion

A mass bouncing up and down on the end of a spring undergoes vibrational motion. The motion of any system whose acceleration is proportional to the negative of displacement is termed **simple harmonic motion** (SHM), i.e. $F = ma = -kx$. Certain definitions pertain to SHM:

- A complete vibration is one down and up motion.
- The time for one complete vibration is the **period,** measured in seconds.
- The **frequency** is the number of complete vibrations per second and is defined as the reciprocal of the period. Its units are cycles/second or hertz (Hz).
- The **amplitude** is the absolute value of the distance from the maximum vertical displacement to the central point of the motion, that is, the greatest distance up or down the mass moves from its initial position.

The equation relating to period, the mass, and the spring constant is $T = 2\pi\sqrt{m/k}$. This relationship gives the period in seconds.

The relation of SHM to circular motion

Aspects of SHM can be visualized by looking at its relationship to uniform circular motion. Imagine a pencil taped vertically to a horizontal turntable. View the rotating pencil from the side of the turntable. As the turntable rotates with uniform circular motion, the pencil moves back and forth with simple harmonic motion. Figure 1-32(a) illustrates P as the point on the rim of the turntable—the position of the pencil. Point P' indicates the apparent position of the pencil when viewing only the x component. The acceleration vector and vector components are shown in Figure 1-32(b).

Figure 1-32 The relationship between circular motion and SHM.

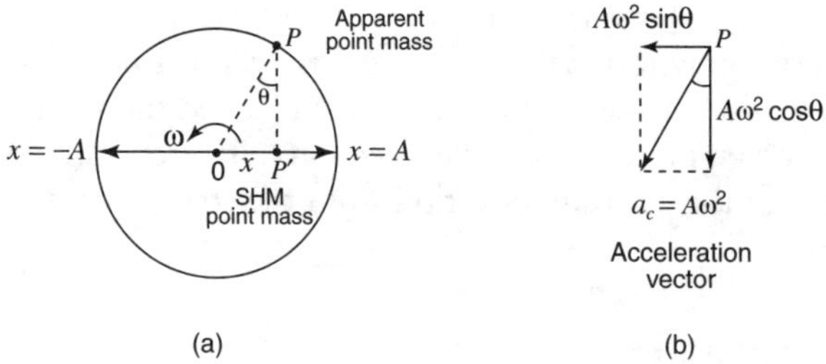

The following is proof of the relationship between SHM and one component of uniform circular motion. This component of motion is that observed by looking at circular motion from the side. The maximum displacement of the component of the uniform circular motion is the radius of the circle *(A)*. Substitute the radius of the circle *(A)* into the equations for angular velocity and angular acceleration to obtain $v = r\omega = A\omega$ and $a = v^2/r = r\omega^2 = A\omega^2$. The horizontal component of this acceleration is $a = -A\omega^2 \sin\theta = -\omega^2 x$, using $x = A\sin\theta$ as shown in Figure 1-32. Because the acceleration is proportional to the displacement, the point rotating with uniform circular motion undergoes SHM when only one component of the motion is considered.

The simple pendulum

The **simple pendulum** is the idealized model of a mass swinging on the end of a massless string. For small arcs of swing of less than 15 degrees, the motion of the pendulum approximates SHM. The period of the pendulum is given by $T = 2\pi\sqrt{l/g}$, where l is the length of the pendulum and g is the acceleration due to gravity. Notice that the period of a pendulum is *not* dependent upon the pendulum's mass.

SHM energy

The potential energy of a Hooke's law spring is $P.E. = (1/2)\,kx^2$. The total energy is the sum of the kinetic and potential energies at any time and is conserved.

Fluids

A **fluid** is a substance that cannot maintain its own shape but takes the shape of its container. Fluid laws assume idealized fluids that cannot be compressed.

Density and pressure

The **density** (ρ) of a substance of uniform composition is its mass per unit volume: $\rho = m/V$. In the SI system, density is measured in units of kilograms per cubic metre.

Imagine an upright cylindrical beaker filled with a fluid. The fluid exerts a force on the bottom of the container due to its weight. **Pressure** is defined as the force per unit area: $\mathbf{P} = \mathbf{F}/\boldsymbol{A}$, or in terms of magnitude, $P = mg/A$, where mg is the weight of the fluid. The SI unit of pressure is N/m^2, called a pascal. The pressure at the bottom of a fluid can be expressed in terms of the density (ρ) and height *(h)* of the fluid:

$$P = \frac{mg}{A} = \frac{\rho Vg}{A} = \frac{\rho hAg}{A}$$

or $P = \rho hg$. The pressure at any point in a fluid acts equally in all directions. This concept is sometimes called the **basic law of fluid pressure.**

Pascal's principle

Pascal's principle may be stated thus: The pressure applied at one point in an enclosed fluid under equilibrium conditions is transmitted equally to all parts of the fluid. This rule is utilized in hydraulic systems. In Figure 1-33, a push on a cylindrical piston at point *a* lifts an object at point *b*.

Figure 1-33 Pascal's principle is used to easily lift a car.

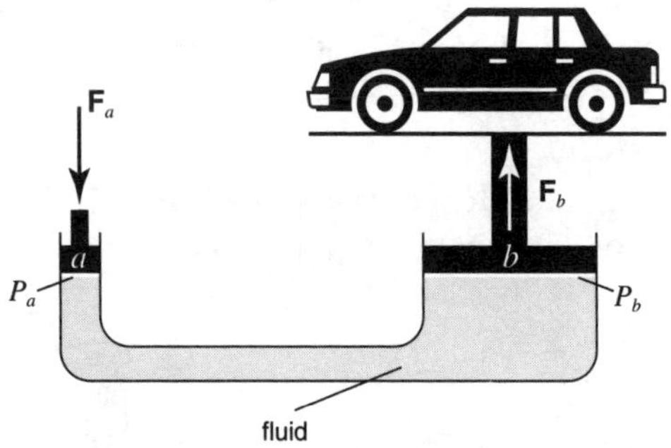

Let the subscripts *a* and *b* denote the quantities at each piston. The pressures are equal; therefore, $P_a = P_b$. Substitute the expression for pressure in terms of force and area to obtain $F_a / A_a = (F_b / A_b)$. Substitute πr^2 for the area of a circle, simplify, and solve for F_b: $F_b = (F_a)\left(r_b^2 / r_a^2\right)$. Because the force exerted at point *a* is multiplied by the square of the ratio of the radii and $r_b > r_a$, a modest force on the small piston *a* can lift a relatively larger weight on piston *b*.

Archimedes' principle

Water commonly provides partial support for any object placed in it. The upward force on an object placed in a fluid is called the **buoyant force.** According to **Archimedes' principle,** the magnitude of a buoyant force on a completely or partially submerged object always equals the weight of the fluid displaced by the object.

Archimedes' principle can be verified by a nonmathematical argument. Consider the cubic volume of water in the container of water shown in Figure 1-34. This volume is in equilibrium with the forces acting on it, which are the weight and the buoyant force; therefore, the downward force of the weight (**W**) must be balanced by the upward buoyant force (**B**), which is provided by the rest of the water in the container.

Figure 1-34 Weight is balanced by buoyant force within a volume of water.

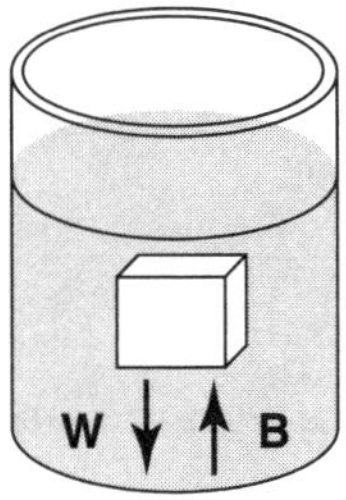

If a solid floats partially submerged in a liquid, the volume of liquid displaced is less than the volume of the solid. Comparing the density of the solid and the liquid in which it floats leads to an interesting result. The formulas for density are $D_s = m_s / V_s$ and $D_l = m_l / V_l$, where D is the density, V is the volume, m is the mass, and the subscripts s and l refer to quantities associated with the solid and the liquid respectively. Solving for the masses leads to $m_s = D_s V_s$ and $m_l = D_l V_l$. According to Archimedes' principle, the weights of the solid and the displaced liquid are equal. Because the weights are simply mass times a constant (g), the masses must be equal also; therefore, $D_s V_s = D_l V_l$ or $D_s / D_l = V_l / V_s$. Now, $V = Ah$, where A is the cross-sectional area and h is the height. For a solid floating in liquid, $A_l = A_s$ and h_s is the height of the solid that is submerged, h_{sub}. With these substitutions, the above relationship becomes $D_s / D_l = h_{sub} / h_s$; therefore, the fractional part of the solid that is submerged is equal to the ratio of the density of the solid to the density of the surrounding liquid in which it floats. For example, about 90 per cent of an iceberg is beneath the surface of sea water because the density of ice is about nine-tenths that of sea water.

Bernoulli's equation

Imagine a fluid flowing through a section of pipe with one end having a smaller cross-sectional area than the pipe at the other end. The flow of liquids is very complex; therefore, this discussion will assume conditions of the smooth flow of an incompressible fluid through walls with no drag. The velocity of the fluid in the constricted end must be greater than the velocity at the larger end if steady flow is maintained; that is, the volume passing per time is the same at all points. Swiftly moving fluids exert less pressure than slowly moving fluids. **Bernoulli's equation** applies conservation of energy to formalize this observation: $P + (1/2)\rho v^2 + \rho gh$ = a constant. The equation states that the sum of the pressure *(P)*, the kinetic energy per unit volume, and the potential energy per unit volume have the same value throughout the pipe.

Chapter Checkout

1. True or False: Two objects with the same mass also always have the same weight.
2. A skier starts at the top of a jump from rest. The height from the starting point to the lip of the ramp is 30 m.
   a. Assuming the ramp is frictionless, how fast will the skier be going when she leaves the ramp?
   b. If the lip of the ramp is inclined at 60° to the horizontal, how long will she be in the air? Assume the landing surface is horizontal and at the same level of the launch point.
3. Two cylinders of equal radii and equal masses rotate about their central axes. One is a thin cylindrical shell, and the other is solid.
   a. If they have the same constant kinetic energy, which spins faster?
   b. If, instead, they spin at the same rate, which has more kinetic energy?
4. The pilot of a plane flying parallel to the ground at a height of 5000 m wishes to drop a bomb on a target below. If the plane is flying at 200 km/h, how far from the target must the bomb be released? Assume the bomb is not subject to air resistance and is released with no initial vertical velocity.
5. A simple pendulum has a bob with a mass of 150 g and swings with a period of 12 s. If the mass of the bob is doubled, how much does the period change?

Answers: 1. False **2. a.** 24.5 m/s **b.** 4.24 s **3. a.** The solid cylinder **b**. The shell **4.** 1757 m **5.** None

Chapter 2
WAVES AND SOUND

CHAPTER CHECK-IN

- ❑ Learning about transverse and longitudinal waves
- ❑ Understanding how waves interfere
- ❑ Understanding the wave properties of sound

The easiest wave to visualize is a water wave. When a pebble is dropped in a calm pool of water, ripples travel out from the point where the pebble enters the water. The disturbance travels out from the centre of the pattern, but the water does not travel with the wave. Mechanical waves—such as water waves, waves on a rope, waves in a spring, and sound waves—have two general characteristics:

- A disturbance is in some identifiable medium.
- Energy is transmitted from place to place, but the medium does not travel between two places.

Wave Motion

For the sake of simplicity, idealized one-dimensional waves on a rope and two-dimensional water surface waves with no friction-like forces provide the wave model. For ease of analysis, a pulse that is a single short wave will be used to illustrate wave characteristics that also hold true for more complex waves.

Transverse and longitudinal waves

On the left side of Figure 2-1, a pulse travels on a string. As the pulse passes point *P* on the string, the point moves up and then back to the equilibrium position. Each segment of the rope moves only perpendicular to the motion of the wave. This type of traveling wave is called a **transverse wave.**

Figure 2-1 Transverse (a) and longitudinal (b) waves.

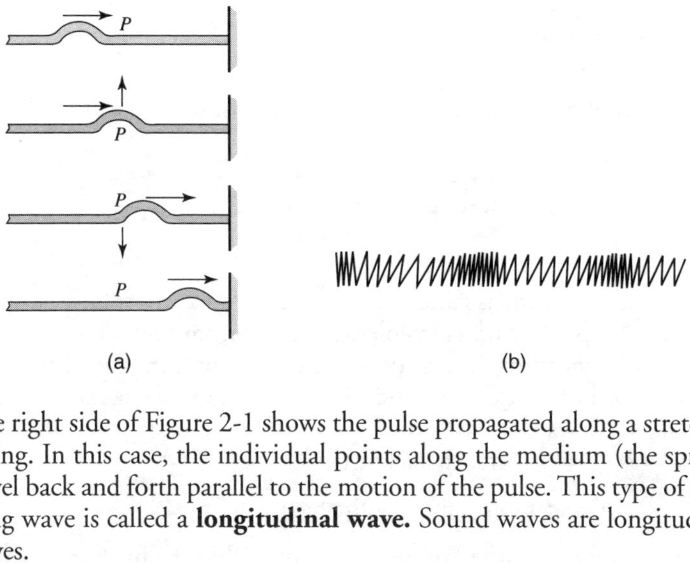

The right side of Figure 2-1 shows the pulse propagated along a stretched spring. In this case, the individual points along the medium (the spring) travel back and forth parallel to the motion of the pulse. This type of traveling wave is called a **longitudinal wave.** Sound waves are longitudinal waves.

Wave characteristics

Important basic characteristics of waves are **wavelength, amplitude, period,** and **frequency.** Wavelength is the length of the repeating wave shape. Amplitude is the maximum displacement of the particles of the medium, which is determined by the energy of the wave. Figure 2-2 illustrates the wavelength (represented by λ, the Greek letter lambda) and the amplitude (by *A*) for both transverse and longitudinal waves.

Figure 2-2 Wavelength λ and amplitude A for transverse and longitudinal waves.

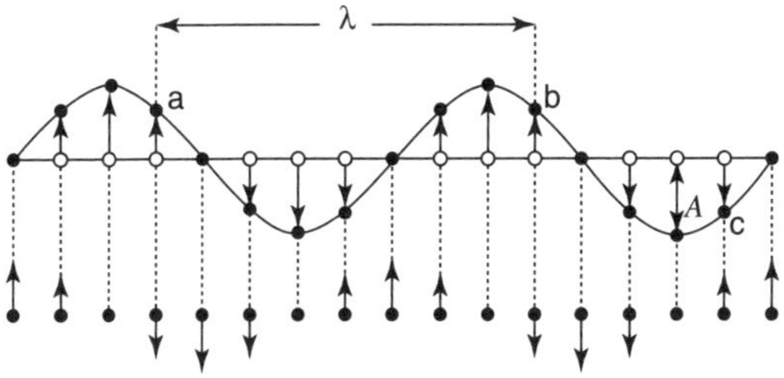

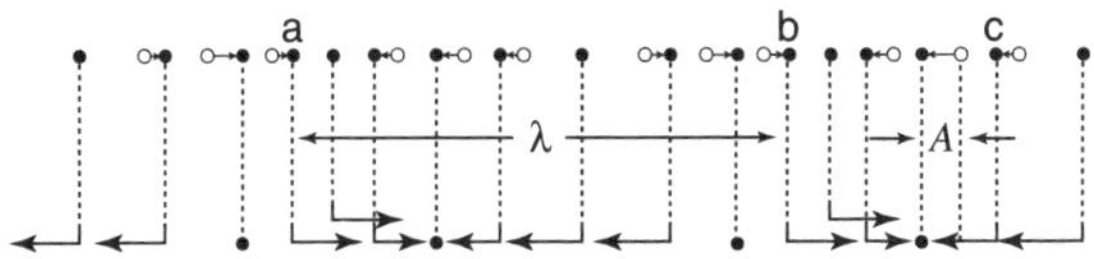

The period (T) is the time for one wave to pass a given point. Period is measured in seconds. Frequency of the wave (f) is the number of waves passing a given point in a unit of time. Frequency is measured in cycles per second or the SI unit of hertz (Hz) with the dimensions of sec^{-1}. For example, a wave generated at 60 cycles per second has a frequency of 60 Hz and can be expressed as 60/s. Frequency is the reciprocal of the period:

$$f = \frac{1}{T}$$

From the definition of velocity as distance/time (distance divided by time)—for all types of waves—the velocity is given by the following:

$$v = \frac{\lambda}{T}$$

This equation states that the wave will advance the distance of one wavelength in the time of one period of vibration. Because frequency is the reciprocal of period, velocity is also $v = \lambda f$. The velocity is dependent upon the characteristics of the medium carrying the wave.

Superposition principle

If two waves pass through the same region of space, they combine by a process called superposition. The **superposition principle** is that the resultant wave formed by the simultaneous influence of two or more waves is the vector sum of the displacements due to each wave acting independently. As shown in Figure 2-3(a), if two pulses of the same size and shape on the same side of the rope arrive at a given point at the same time, they will—for an instant—combine to form a pulse that is twice the size of each of the individual pulses. This is called **constructive interference.** Figure 2-3(b) shows what happens if the same two pulses are on opposite sides of the string. In this case, the two pulses will momentarily cancel each other out. This is called **destructive interference.**

Figure 2-3 Constructive interference (a) and destructive interference (b).

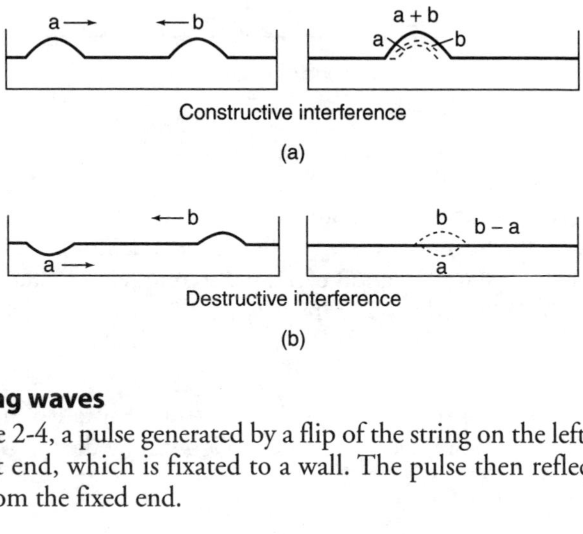

Standing waves

In Figure 2-4, a pulse generated by a flip of the string on the left travels to the right end, which is fixated to a wall. The pulse then reflects upside down from the fixed end.

Figure 2-4 A wave pulse reflects from a fixed barrier.

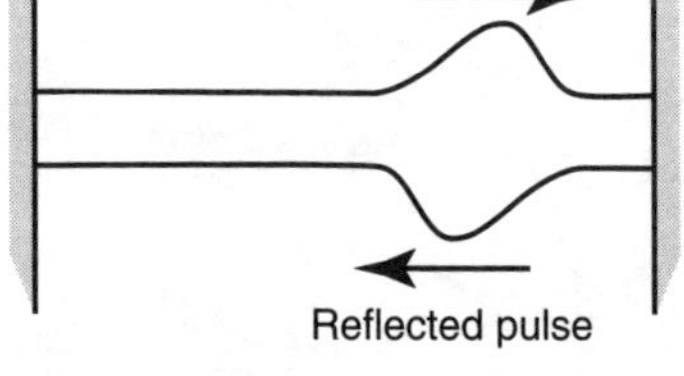

Now, suppose that pulses are sent along the string at regular time intervals. The reflected pulse traveling to the left adds to the original pulse traveling to the right toward the wall. **Standing waves** are produced by the superposition of these similar but inverted pulses that are traveling in opposite directions. Figure 2-5 shows successive time frames as the pulses pass through each other.

Figure 2-5 Incident and reflected waves interfere (a) to create a standing wave (b).

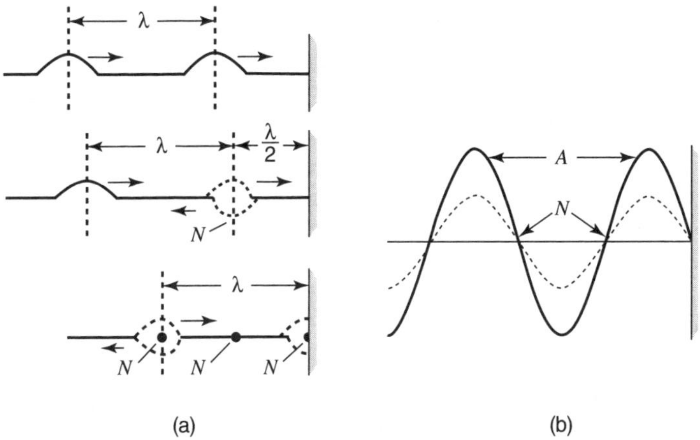

Note that certain points do not move. At these points, there is always a displacement in one direction from one pulse that is canceled by an equal and opposite displacement from a reflected pulse. These points are called **nodal** points *(N)*. For waves, which are pulses that alternate direction of displacement, halfway between the nodal points are segments that move up and down with a maximum displacement of twice the amplitude of the original wave. Such points are **antinodes** *(A)*. The wavelength of the component waves (original and reflected) is twice the distance between adjacent nodes in the standing (resultant) wave.

Sound

Sound waves are produced by a vibrating body. The vibrating object moves in one direction and compresses the air directly in front of it. As the vibrating object moves in the opposite direction, the pressure on the air is lessened so that an expansion, or rarefaction, of air molecules occurs. One compression and one rarefaction make up one longitudinal wave. The vibrating air molecules move back and forth parallel to the direction of motion of the wave, receiving energy from adjacent molecules nearer the source and passing the energy to adjacent molecules farther from the source.

Intensity and pitch

The **pitch** of a sound depends on the frequency of the tone that the ear receives. High notes are produced by an object that is vibrating a greater number of times per second than for a low note.

The **intensity** of a sound is the amount of energy crossing a unit area in unit time or the power flowing through the unit area. The SI unit is watts per square metre. The **loudness** of the sound depends upon the subjective effect of intensity of sound waves on the human ear. In general, a more intense sound is also louder, but the ear does not respond similarly at all frequencies so that two tones of the same intensity but with different pitches may appear to have different loudness. The intensity of the threshold of hearing (I_0), which is the intensity that can be barely heard by a normal person, is about 10^{-12} watt/m^2 when measured by acoustical devices. The relation between loudness and intensity is nearly logarithmic. The intensity level of sound is measured in **decibels** and is given by the equation $\beta = 10 \log I/I_0$, where β (the Greek letter beta) is the intensity in decibels, I is the sound intensity, and I_0 is the intensity of the threshold of hearing. For example, normal conversation is about 60 decibels, and a power saw is about 110 decibels.

Doppler effect

When a siren approaches, the pitch is high, and when it passes, the pitch drops. As a moving sound source approaches a listener, the sound waves are closer together, as shown in Figure 2-6, causing an increase in the frequency of the sound heard. As the source passes the listener, the waves spread out, and the observed frequency lowers.

Figure 2-6 The Doppler effect.

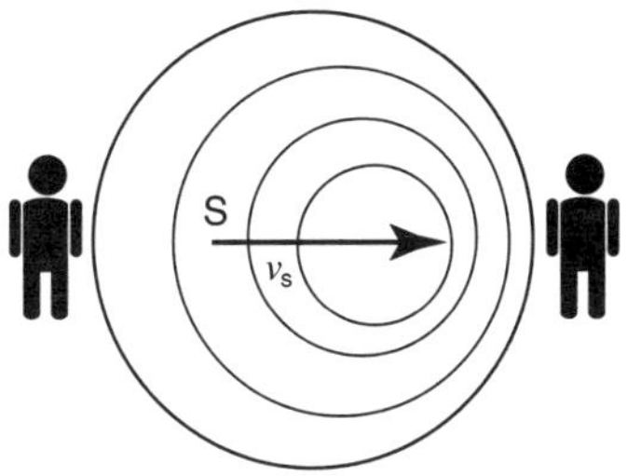

This change in observed frequency due to relative motion is called the **Doppler effect.** The equation for a stationary observer and moving source is

$$f' = f\left(\frac{v}{v \pm v_s}\right) \text{(moving source)}$$

where f' is the frequency heard by the observer, f is the source frequency, v is the speed of sound in air, and v_s is the speed of the source.

Forced vibrations and resonance

The tuning fork is a useful instrument for investigating sound because it vibrates at only one frequency, in contrast to most musical instruments that produce several different frequencies simultaneously. A struck tuning fork vibrates at a natural frequency that depends upon the fork's manufacture—the dimensions and the material from which it is made. If the stem of a vibrating tuning fork is set on a table top, the tone becomes louder because the fork forces the table top to vibrate. Because the table top has a larger vibrating area, the sound is more intense. This principle of forced vibrations is applied in most musical instruments by using a part of the instrument, such as the piano sounding board, to intensify the sound.

Imagine two matched tuning forks with the same frequency mounted on sounding boxes. As shown in Figure 2-7, the vibrating air column set up by one tuning fork will cause the other tuning fork to vibrate weakly. This action is called **resonance** or **sympathetic vibration.** Resonance occurs when the natural vibration rates of two objects are the same or when one has a natural vibration rate that is a multiple of the other. The requirement that the two objects have the same natural frequency (or multiple thereof) can be demonstrated by a violin and a tuning fork that vibrates at the pitch of one of the violin strings. First, set small pieces of paper on each violin string. Then hold the tuning fork very near the violin strings. The small paper will fall off the string that has the same natural frequency as the tuning fork because the string experiences a weak sympathetic vibration.

Figure 2-7 Sympathetic vibration between two tuning forks.

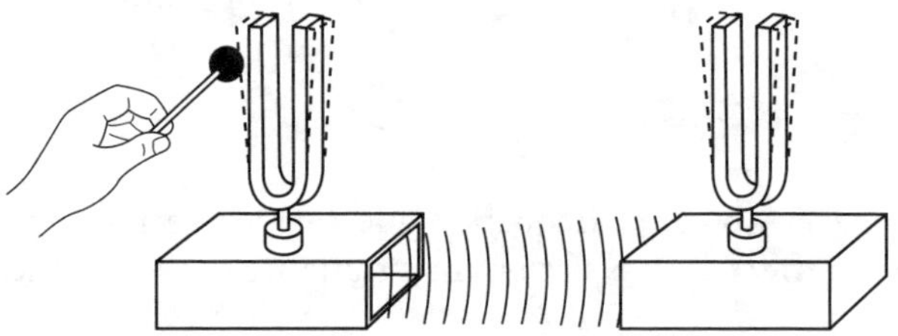

Beats

The discussion of standing waves analyzed the superposition of waves with the same frequency. A different interference effect occurs when two waves with slightly different frequencies are heard at the same time. The top graph in Figure 2-8 represents the individual waves of two slightly different frequencies. The bottom graph shows the resultant wave. At time t_a, the two waves destructively interfere (cancel each other out). At a later time (t_b), the waves constructively interfere because the amplitudes are both in the same direction. A listener will hear the alternating loudness, known as **beats.** The number of beats per second, called the beat frequency, equals the difference between the frequencies of the two individual waves. To tune an instrument accurately, a musician listens carefully and adjusts her instrument to eliminate beats between the instrument and a given pitch.

Figure 2-8 Beats are created by the interference of two waves with different frequencies.

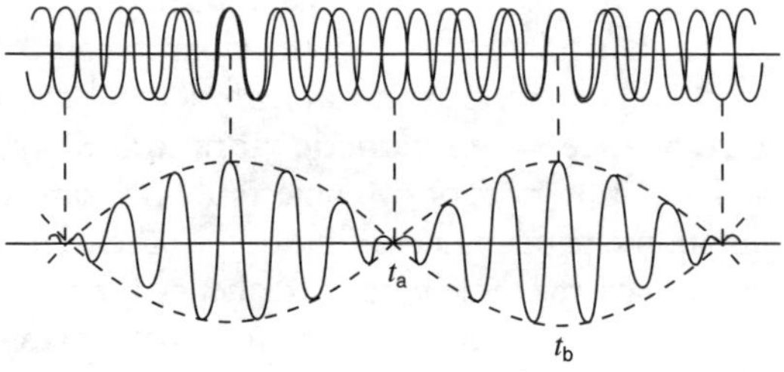

Chapter Checkout

1. True or False: Sound waves can be either transverse or longitudinal.
2. Two trains are approaching each other from opposite directions. The *Wabash Cannonball* is traveling at 200 km/h, while the *City of New Orleans* is traveling at 160 km/h.
   a. At the same time, both trains whistle: *Wabash Cannonball* at 4800 Hz and *New Orleans* at 7200 Hz. What frequency does each train hear from the other?
   b. After they pass each other, they blow their whistles. What frequencies do they hear now? Assume the speed of sound is 340 m/s.
3. If a standing wave has frequency 30 Hz and velocity 4.5 m/s, how far apart are the nodal points?
4. True or False: A buoy moves only up and down when a transverse water wave passes by, but it does not move along with the wave.

Answers: 1. False **2. a.** *WC* hears 10200 Hz; *NO* hears 6800 Hz. **b.** *WC* hears 5563 Hz; *NO* hears 3863 Hz. **3.** 75 cm **4.** True

Chapter 3
THERMODYNAMICS

CHAPTER CHECK-IN

- ❑ Understanding temperature and thermal expansion and contraction
- ❑ Applying the ideal gas law
- ❑ Learning about heat
- ❑ Understanding the first and second laws of thermodynamics

Thermodynamics is the field of physics that successfully explains the properties of matter visible in our everyday, macroscopic world and the correlation between those properties and the mechanics of atoms and molecules.

Temperature

Temperature measures how hot or cold a body is with respect to a standard object. To discuss temperature changes, two basic concepts are important: thermal contact and thermal equilibrium. Two objects are in **thermal contact** if they can affect each other's temperature. **Thermal equilibrium** exists when two objects in thermal contact *no longer* affect each other's temperature. For example, if a carton of milk from the refrigerator is set on the kitchen countertop, the two objects are in thermal contact. After several hours, their temperatures are the same, and they are then in thermal equilibrium.

Thermometers and temperature scales

The sense of touch provides some indication of the temperature of an object but is unreliable. For example, the metal shelf in the refrigerator feels colder than the food sitting on the shelf, even though they are in thermal equilibrium. The metal feels colder because the metal conducts the heat from your hand more efficiently.

Thermometers are instruments that define and measure the temperature of a system. The common thermometer consists of a volume of mercury that expands into a capillary tube when heated. When the thermometer is in thermal equilibrium with an object, the temperature can be read from the thermometer scale.

Three temperature scales are commonly used: Celsius, Fahrenheit, and Kelvin (also called *absolute*). Comparisons of the Celsius and Fahrenheit thermometers are shown in Figure 3-1.

Figure 3-1 Comparison of Celsius and Fahrenheit thermometers.

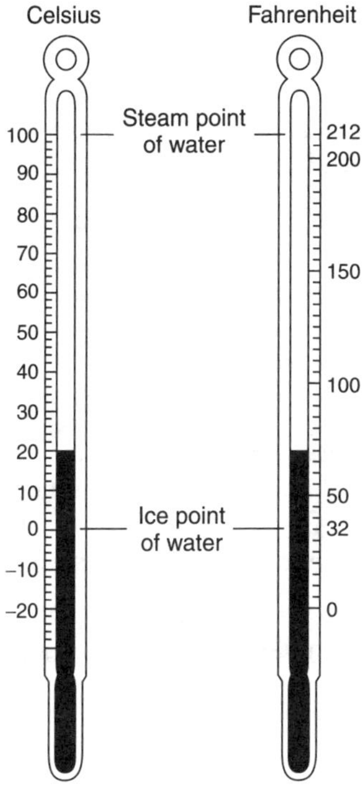

On the Celsius scale, the ice point is 0, and the steam point is 100. The interval between these temperatures is divided into 100 equal parts called **degrees.** As shown in Figure 3-1, on the Fahrenheit scale, the ice point is 32 degrees, and the steam point is 212 degrees. The interval between these temperatures is divided into 180 equal parts. The following equations relate temperature in Celsius (C) and Fahrenheit (F):

$$T_C = \frac{5}{9}(T_F - 32°) \text{ and } T_F = \frac{9}{5}T_C + 32°$$

The Kelvin scale (K) has degrees of the same size as the Celsius scale, but the zero is shifted to the **triple point of water.** The triple point of water exists when water within a closed vessel is in equilibrium in all three states: ice, water, and vapour. This point is defined as 273.16 Kelvin and equals .01 degrees Celsius; therefore, to convert Celsius to Kelvin, simply add 273.15. Note that because the degrees are the same in the two scales, temperature differences are the same in either Celsius or Kelvin.

Thermal expansion of solids and liquids

A mercury thermometer utilizes **thermal expansion:** the phenomenon that most substances increase in volume as their temperature increases. A rod that is heated will change in length (ΔL) according to $\Delta L = \alpha L_o \Delta T$, where L_o is the original length and ΔT (delta T) is the change in temperature. The constant α (Greek letter alpha) is the average coefficient of linear expansion. This value is found in tables of coefficients for different materials and is measured in units of (degrees C)$^{-1}$.

Not only does length change with a change in temperature, but area and volume change also. Thus, $\Delta A = \gamma A_o \Delta T$, where ΔA is the change in the original area A_o. The Greek letter gamma (γ) is the average coefficient of area expansion, which equals 2α. For change in volume, $\Delta V = \beta V_o \Delta T$, where ΔV is the change in the original volume V_o. The Greek letter beta (β) is the average coefficient of volume expansion, which is equal to 3α.

Example 1: As an example of the application of these equations, consider heating a steel washer. What will be the area of the washer hole with original cross-sectional area of 10 mm^2 if the steel has $\alpha = 1.1 \times 10^{-5}$ per °C and is heated from 20 degrees C to 70 degrees C?

Solution: The hole will expand the same as a piece of the material having the same dimensions. The equation for increase in area leads to the following:

$$\Delta A = 2\alpha A_o \Delta T$$

$$= [2 \times 1.1 \times 10^{-5} (°\text{C})^{-1}][10 \text{ mm}^2][50 °\text{C}]$$

$$= 0.011 \text{ mm}^2$$

Therefore, the new area of the hole will be 10.011 mm^2.

Water is an exception to the usual increase in volume with increasing temperature. Note in Figure 3-2 that the maximum density of water occurs at 4 degrees Celsius.

Figure 3-2 The density of water changes as the temperature changes.

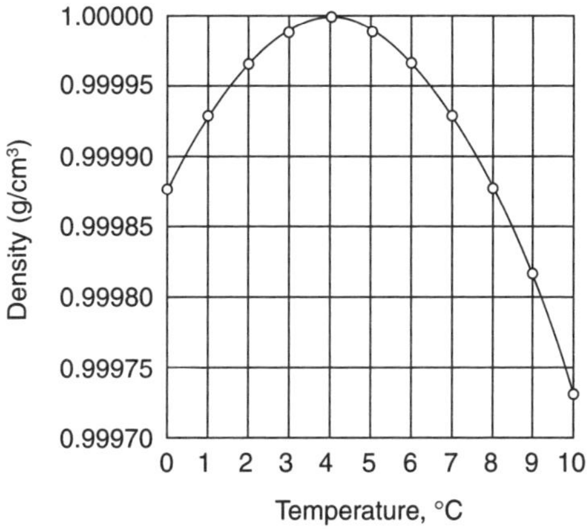

This characteristic of water explains why a lake freezes at the surface. To see this, imagine that the air cools from 10 degrees Celsius to 5 degrees Celsius. The surface water in equilibrium with the air at these temperatures is denser than the slightly warmer water below it; therefore, the colder water sinks and warmer water from below comes to the surface. This occurs until the air temperature decreases to below 4 degrees when the surface water is less dense than the deeper water of about 4 degrees; then, the mixing ceases. As the temperature of the air continues to fall, the surface water freezes. The less dense ice remains on top of the water. Under these conditions, life near the bottom of the lake can continue to survive because only the water at or near the surface is frozen. Life on earth might have evolved quite differently if a pool of water froze from the bottom up.

Development of the Ideal Gas Law

The pressure, volume, temperature, and amount of an ideal gas are related by one equation that was derived through the experimental work of several individuals, especially Robert Boyle, Jacques A. C. Charles, and Joseph Gay-Lussac. An **ideal gas** consists of identical, infinitesimally small particles that only interact occasionally like elastic billiard balls. Real gases act much like ideal gases at the usual temperatures and pressures found on the earth's surface. The gases in the sun are not ideal gases due to the high temperature and pressures found there.

Boyle's law

If a gas is compressed while keeping the temperature constant, the pressure varies inversely with the volume. Hence, **Boyle's law** can be stated thus: The product of the pressure *(P)* and its corresponding volume *(V)* is a constant. Mathematically, PV = constant. Or, if P is the original pressure, V is the original volume, P' represents the new pressure, and V' the new volume, the relationship is

$$V' = V\frac{P}{P'}$$

Charles/Gay-Lussac law

The **Charles/Gay-Lussac law** denotes that for a constant pressure, the volume of a gas is directly proportional to the Kelvin temperature. In equation form, V = (constant)T. Or if V is the original volume, T the original Kelvin temperature, V' the new volume, and T' the new Kelvin temperature, the relationship is

$$\frac{V}{T} = \frac{V'}{T'}$$

Boyle's law and the Charles/Gay-Lussac law can be combined: PV = (constant)T. The volume increases when the mass *(m)* of gas increases as, for example, pumping more gas into a tire; therefore, the volume of the gas is also directly related to the mass of the gas and PV = (constant)mT.

Definition of a mole

The proportionality constant of the previous equation is the same for all gases if the amount of gas is measured in **moles** rather in terms of mass. The number of moles *(n)* of gas is the ratio of the mass *(m)* and the molecular or **atomic** mass *(M)* expressed in grams per mole:

$$n = \frac{m}{M}$$

The mole of pure substance contains a mass in grams equal to the molecular mass or atomic mass of the substance. For example, lead has an atomic mass of 207 g/mole, or 207 g of lead is 1 mole of lead.

The ideal gas law

Incorporating Boyle's law, the Charles/Gay-Lussac law, and the definition of a mole into one expression yields the **ideal gas law** $PV = nRT$, where R is the **universal gas constant** with the value of R = 8.31 J/mole-degree × K in SI units, where pressure is expressed in **N/m²** (pascals), volume is in cubic metres, and temperature is in degrees Kelvin.

If the temperature, pressure, and volume change for a given number of moles of gas, the formula is

$$\frac{PV}{T} = \frac{P'V'}{T'}$$

where unprimed variables refer to one set of conditions and the primed variables refer to another. Frequently, a set of conditions of the temperature, pressure, and volume of a gas are compared to standard temperature and pressure (STP). **Standard pressure** is 1 atmosphere, and **standard temperature** is 0 degrees Celsius (approximately 273 degrees Kelvin).

Avogadro's number

Amadeo Avogadro (1776–1856) stated that one mole of any gas at standard pressure and temperature contains the same number of molecules. The value called **Avogadro's number** is $N = 6.02 \times 10^{23}$ molecules/mole. The ideal gas law can be written in terms of Avogadro's number as $PV = NkT$, where k, called the Boltzmann's constant, has the value $k = 1.38 \times 10^{-23}$ J/K. One mole of any gas at standard temperature and pressure (STP) occupies a **standard volume** of 22.4 litres.

The kinetic theory of gases

Consider a gas with the four following idealized characteristics:

- It is in thermal equilibrium with its container.
- The gas molecules collide elastically with other molecules and the walls of the vessel.
- The molecules are separated by distances that are large compared to their diameters.
- The net velocity of all the gas molecules must be zero so that, on the average, as many molecules are moving in one direction as in another.

This model of a gas as a collection of molecules in constant motion undergoing elastic collisions according to Newtonian laws is the **kinetic theory of gases.**

From Newtonian mechanics, the pressure on the wall *(P)* may be derived in terms of the average kinetic energy of the gas molecules:

$$P = \frac{2}{3}\left(\frac{N}{V}\right)\left(\frac{1}{2} m_o v_{avg}^2\right)$$

The result shows that the pressure is proportional to the number of molecules per unit volume *(N/V)* and to the average linear kinetic energy of the molecules. Using this formula and the ideal gas law, the relationship between temperature and average linear kinetic energy can be found:

$$\frac{3}{2}kT = \frac{1}{2}m_o v_{avg}^2$$

where *k* is again Boltzmann's constant; therefore, the average kinetic energy of gas molecules is directly proportional to the temperature of the gas in degrees Kelvin. Temperature is a direct measure of the average molecular kinetic energy for an ideal gas.

These results seem intuitively defensible. If the temperature rises, the gas molecules move at greater speeds. If the volume remains unchanged, the hotter molecules would be expected to hit the walls more often than cooler ones, resulting in an increase in pressure. These significant relationships link the motions of the gas molecules in the subatomic world to their characteristics observed in the macroscopic world.

Heat

Heat and temperature are different physical quantities. When two objects with different temperatures are in contact with each other, heat flows from the hotter system to the colder one. Heat is a measure of energy. An increase or decrease in mechanical energy in a system always accompanies an equal decrease or increase of heat, respectively.

Heat capacity and specific heat

The **heat capacity** of a body is the amount of heat energy necessary to raise the temperature of an object by one degree. Imagine blocks of the same mass made of different metals (see Figure 3-3). The blocks have bases of the same cross section but different heights because the densities are different. After being heated in the oven to the same temperature, the blocks are placed on a large piece of ice. Some of the blocks will melt further into the ice than others as the result of their different abilities to absorb or give out different amounts of heat even though they have the same mass and undergo the same change in temperature. The blocks differ in heat capacity. Differences in heat capacity may also be a result of different masses and different temperature changes.

Figure 3-3 Hot blocks of different metals, at the same temperature, melt different amounts of ice.

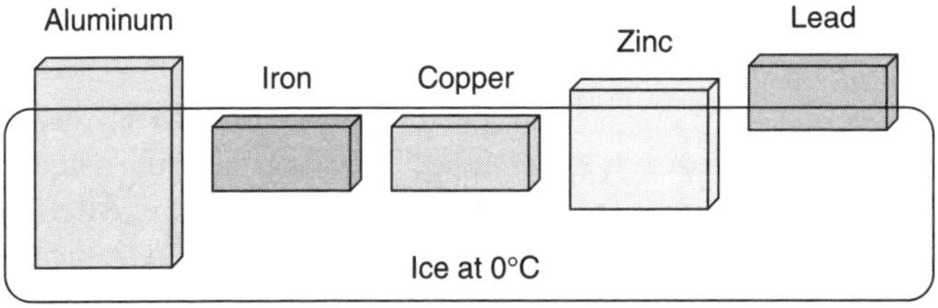

The heat capacity *(C)* per unit mass *(m)* is called **specific heat** (*c*):

$$c = \frac{C}{m} = \frac{Q/\Delta T}{m}$$

where Q units of heat are added to m kg of a substance, changing the temperature by ΔT. The specific heats have been determined for many materials and can be found in tables.

Mechanical equivalent of heat

The **calorie** is defined as the amount of energy required to raise 1 gram of water 1 degree. (This energy is slightly dependent upon the temperature of the water, so the temperature change is usually defined from 14.5 degrees to 15.5 degrees Celsius.) The **kilocalorie** is the amount of heat energy needed to raise 1 kilogram of water by 1 degree Celsius. (Food calories are kilocalories.) In SI units, the calorie equals 4.184 joules. The U.S. engineering unit of heat is the **British thermal unit** (BTU). It is related to the calorie and the joule: BTU = 252 calories = 1.054 kJ. These reversible conversions of heat energy and work are called the **mechanical equivalent of heat.**

Heat transfer

The heat energy *(Q)* transferred into or out of a system is given by $Q = mc\Delta T$. The temperature change is positive for a gain in heat energy and negative for heat removed from the object. When applying this expression in heat exchange problems, assume that the objects in thermal contact are isolated from their surroundings—completely insulated.

Calorimetry

If a substance in a closed container loses heat, then something else in the container gains an equal amount of heat. A **calorimeter** is a device that utilizes the transfer of heat to determine the specific heat of a substance. A known mass of a substance whose specific heat is unknown is heated to a certain temperature and then placed in a container containing a liquid (usually water) of known mass, specific heat, and temperature. After thermal equilibrium is reached, the specific heat of the unknown can be determined.

Example 2: Consider a block of hot metal, with mass m_m and original temperature T_{mo}, which is dropped into a mass of cool water of mass m_w with beginning temperature T_{wo}. If the final temperature is T_f, what is the specific heat of the metal?

Solution: All of the heat lost by the metal is gained by the water because the system is isolated. The heat lost by the unknown is $Q_m = m_m c_m \Delta T_m = m_m c_m (T_{mo} - T_f)$, and the heat gained by the water is $Q_w = m_w c_w \Delta T_w = m_w c_w (T_f - T_{wo})$. The temperature differences have been written so that they are both positive quantities. The final temperature of the water will be greater than its original temperature because it is warming. The final temperature of the metal will be less than its original temperature. The objects attain thermal equilibrium, and so the final temperatures are the same. (The specific heat for water has the value of 1 cal/g.k.)

$$Q_w = Q_m$$

$$m_w c_w (T_f - T_{wo}) = m_m c_m (T_{mo} - T_f)$$

$$C_m = C_w \frac{M_w}{M_m} \frac{(T_f - T_{wo})}{(T_{mo} - T_f)}$$

Latent heat

A **change of phase** occurs when an object changes from one physical state to another. The common **physical states** are solid, liquid, or gas. Some examples of phase changes are from a liquid to a solid (freezing) or from a liquid to a gas (boiling).

The plot shown in Figure 3-4 illustrates the temperature versus heat added to a mass of ice that undergoes phase changes from ice to water and then from water to steam. In regions I, III, and V, the addition of heat energy increases the temperature of the sample; however, in regions II and IV, additional heat does not cause a change in temperature because heat is required to change the state. The heat required for a change of state is called **latent heat** *(L)*: $Q = mL$.

Figure 3-4 Phase changes of water as heat is added.

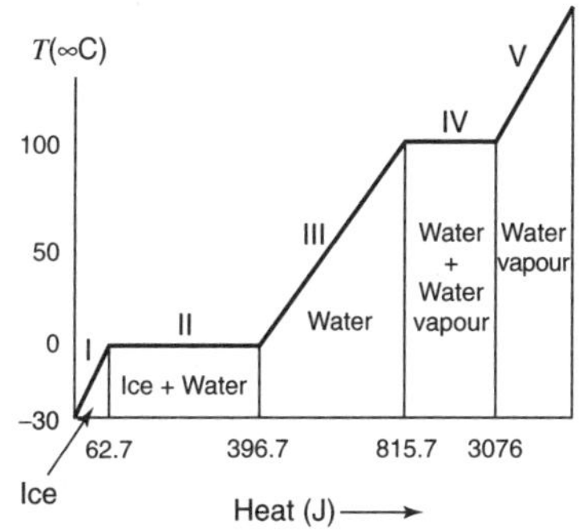

The heat of fusion

The value of latent heat *(L)* depends upon the particular phase change as well as the properties of the substance. The **heat of fusion** is the heat required for a phase change from a solid to a liquid. If the substance is originally in liquid form, the heat of fusion is the heat released when the substance changes from a liquid to a solid. The latent heat of fusion for water at atmospheric pressure is 3.34×10^5 J/kg. The heat required to melt 1 gram of water at 0°C is

$$Q = mL_f = \left(10^{-3}\text{ kg}\right)\left(3.34 \times 10^5\text{ J/kg}\right)$$
$$Q = 334\text{ J}$$

The heat of vaporization

The latent **heat of vaporization** concerns the phase change between the liquid and gaseous states. The heat of vaporization for water is 2.26×10^6 J/kg. The amount of heat necessary to convert 1 gram of water to steam at 100°C is $Q = mL_v = (10^{-3}\text{kg})(2.26 \times 10^6\text{ J/kg}) = 2.26 \times 10^3$ J. Continued addition of heat to steam will cause the steam to be superheated, to attain a higher temperature than 100°C.

Note from the graph in Figure 3-4 that a glass with a mixture of ice and water remains at the temperature of 0°C. Only after all the ice is melted will continued heating increase the temperature of the solution. A similar effect occurs when water and water vapour exist at the same temperature of 100°C until all the water is boiled away.

Methods of heat transfer: Conduction, convection, and radiation

Heat energy can be transferred from one location to another by one of three methods: conduction, convection, and radiation.

The metal handle of an iron skillet placed on a heated burner gets hot by conduction. **Conduction** occurs when the heat travels through the heated solid. The **transfer rate** *(H)* is the ratio of the amount of heat per time transferred from one location in an object to another $H = Q/\Delta t$, where H has units of watts or J/s, when Q is in joules, and Δt is in seconds. The temperature between two parts of the conducting medium—the pan bottom and the handle—must be different for conduction to take place. The formula for heat conduction from one side to another of a slab with thickness L and cross-sectional area A is given by

$$H = \frac{Q}{\Delta t} = kA\frac{(T_2 - T_1)}{L}$$

where the heat flows from T_2 to T_1 and $T_2 > T_1$, as shown in Figure 3-5.

Figure 3-5 Heat flows from regions of higher temperature to lower.

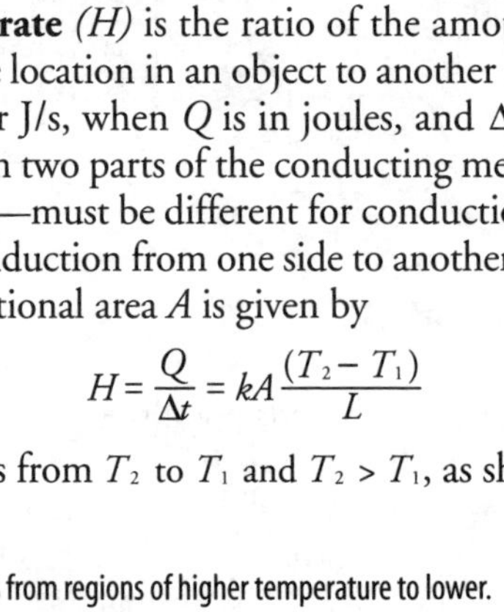

The constant *(k)*, called **thermal conductivity,** is found in tables listing properties of materials. The fact that different materials have different k values explains why the metal shelf of a refrigerator feels colder than the food on it even if both are at thermal equilibrium. The conductivity constant is relatively large for metals, and the metal feels colder because the heat is conducted away from the hand more quickly by metal than by other materials.

Heat transported by the movement of a heated substance is a result of **convection.** The most common example of convection is the warmed mass of air rising from a heater or fire.

The third mechanism for heat transfer is **radiation** in the form of electromagnetic waves. Radiant energy from the sun warms the earth. The rate at which an object emits radiant energy is proportional to the fourth power of its absolute temperature. The **Stefan-Boltzmann law,** which describes the relationship, is written $P = \sigma AeT^4$, where P is the power radiated in watts, σ is a constant equal to $5.6696 \times 10^{-8}\,\mathrm{W/m^2K^4}$, A is the surface area of the object in m^2, T is the absolute temperature, and e is the **emissivity constant,** which varies from 0 to 1 depending upon the properties of the surface.

The thermos bottle, or Dewar flask, is an object that minimizes heat transfer by conduction, convection, and radiation. The flask is constructed of double-walled Pyrex glass with silvered inner walls. The space between the walls is evacuated to reduce heat transfer by conduction and convection. The silvered walls reflect most of the radiant heat to cut heat transfer by radiation. The container is effectively used to store either cold or hot liquids for long periods of time.

The Laws of Thermodynamics

The laws of thermodynamics involve the relations between heat and mechanical, electrical, and other forms of energy or work. The laws are valid only when applied to systems in thermal equilibrium and *not* for systems in the process of rapid change or with complicated states of transition. A system very nearly in equilibrium all the time is called a **reversible** system.

The first law of thermodynamics

The **first law of thermodynamics** is the restatement of conservation of energy. Mathematically, it reads $\Delta Q = \Delta U + \Delta W$, where ΔQ is the heat energy supplied to the system, ΔU is the change in the internal energy, and ΔW is the work done by the system against external forces. It must be emphasized that these quantities are defined in general terms. The internal energy includes not only mechanical energy, but also the rotational and vibrational energy of the molecules, as well as the chemical energy stored in interatomic forces. Work is not only mechanical work but includes other forms, such as work done by electrical currents.

Work

Imagine a system of gas in a cylinder fitted with a piston, as shown in Figure 3-6.

Figure 3-6 A cylinder filled with gas, with a piston.

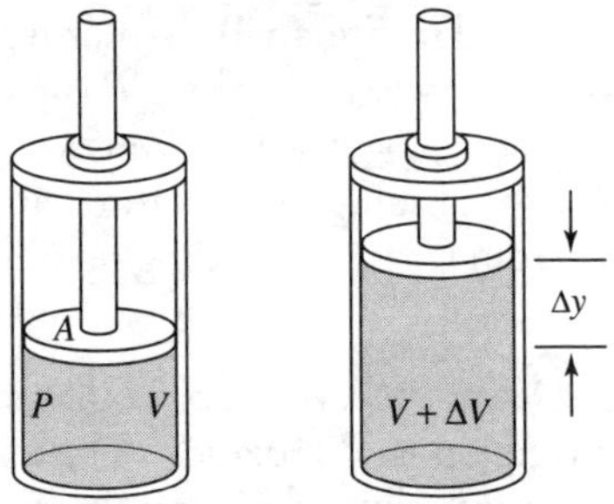

As the gas in the cylinder expands, the force exerted by the gas on the piston is $F = PA$. The piston moves up a distance Δy; therefore, the work done by the gas is $W = F\Delta y = PA\Delta y$, or $W = P\Delta V$ because $A\Delta y$ is the increase in volume (V) of the gas. In general, work done by an expanding gas equals the area under a pressure-volume curve.

Definitions of thermodynamical processes

Four definitions are commonly used to describe system changes in ideal gases where one of the four thermodynamic variables—temperature, volume, pressure, and heat—remains constant. The pressure-volume graphs for these four different processes are shown in Figure 3-7.

The **isobaric** process is shown in Figure 3-7(a), where the pressure of the system remains constant. Both the volume and temperature change. The **isothermal** process is shown in Figure 3-7(b), where the temperature of the system remains constant; therefore, by the ideal gas laws, the product of the volume and the pressure remains constant. An **adiabatic** process is shown in Figure 3-7(c), where there is no heat exchange with the outside world. An **isochoric** process is shown in Figure 3-7(d), where the volume of the system remains constant as the pressure and temperature change.

Figure 3-7 Pressure-volume graphs for thermodynamic processes: (a) isobaric, (b) isothermal, (c) adiabatic, (d) isochoric.

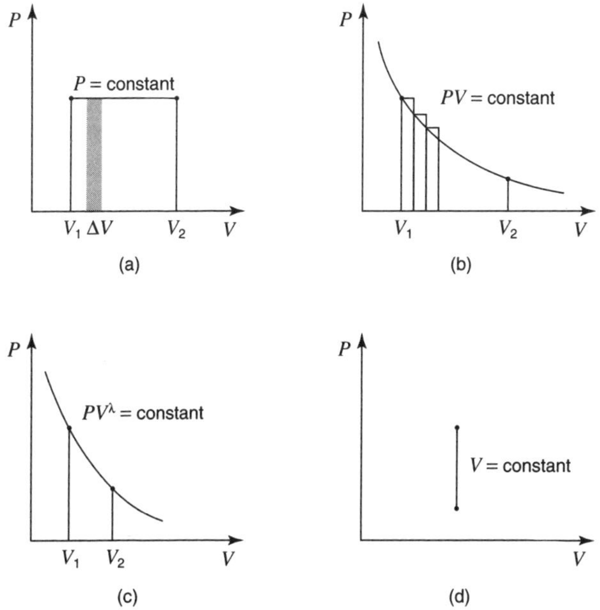

In each case, the work done is the area under the curve. Note that in Figure 3-7(d), the area under the curve is zero; no work is done in the isochoric process.

Carnot cycle

The engineer N. L. Sadi Carnot (1796–1832) first proposed an ideal heat engine that operated through a cycle of reversible isothermal and adiabatic steps. Imagine the engine to be an idealized gas in a cylinder with a fitted piston that supports a load as shown in Figure 3-8. During four steps on one down and upward stroke of the piston, visualize the gas and cylinder sitting first on a heat source (heat is added), then on an insulator (no heat exchange), next on a heat sink (heat is removed), and finally back on the insulator.

Figure 3-8 The Carnot cycle.

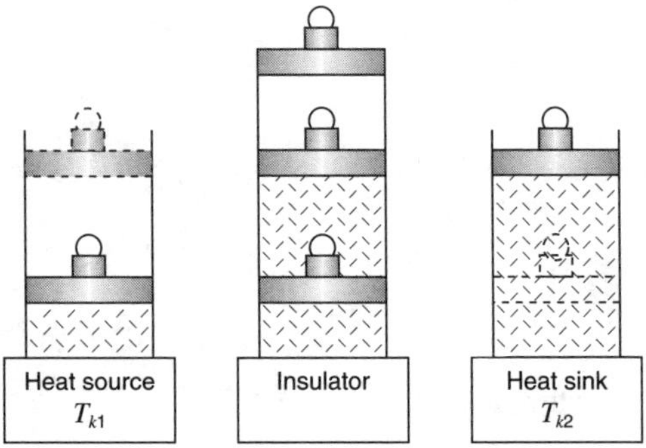

The pressure-volume curve of Figure 3-9 shows the **Carnot cycle.** The gas in the cylinder contains an ideal gas at pressure *(P)*, volume *(V)*, and temperature *(T)*—point A on the curve. The cylinder with gas is set on a heat source and expands isothermally (the temperature remains constant as the pressure decreases and the volume increases) to point B on the graph. During this isothermal expansion, the gas did work lifting a load (or turning a wheel). This work is represented by the area under the A-B curve between V_1 and V_2. Now, the gas and cylinder are placed on an insulator; the gas expands adiabatically (no heat exchange with the outside world) to point C on the curve. More work is done *by the gas on the piston* through this expansion, represented by the area under the B-C curve between V_2 and V_3.

Figure 3-9 P-V graph for the Carnot cycle.

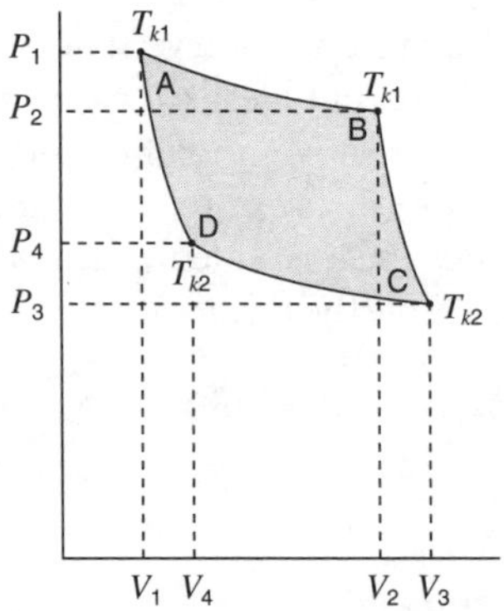

Next, the gas and cylinder are placed on a heat sink. The gas is compressed isothermally and gives up an amount of heat to the heat sink. The conditions at point D describe the gas. For this segment, work is done by the *piston on the gas,* which is represented by the area under the C-D segment of the curve from V_3 to V_4. Finally, the gas and cylinder are placed back on the insulator. The gas is further compressed adiabatically until it returns to the original conditions at point A. Again, for this part of the Carnot cycle, work is done on the gas, which is represented by the area under the D-A segment between V_4 and V_1.

The total work done by the gas on the piston is the area under the ABC segment of the curve; the total work done on the gas is the area under the CDA segment. The difference between these two areas is the shaded portion of the graph. This area represents the work output of the engine. According to the first law of thermodynamics, there is no permanent loss or gain of energy; therefore, the work output of the engine must equal the difference between the heat absorbed from the heat source and that given up to the heat sink.

Consideration of the work output and input leads to the definition of efficiency of an ideal heat engine. If the energy absorbed from the heat source is Q_1 and the heat given up to the heat sink is Q_2, then work output is given by $W_{output} = Q_1 - Q_2$. Efficiency is defined as the ratio of the work output over the work input expressed in per cent, or

$$\text{efficiency} = \frac{W_{output}}{W_{input}} \times 100\%$$

which when expressed in terms of heat is

$$\text{efficiency} = \frac{Q_1 - Q_2}{Q_1} \times 100\%$$

and in terms of temperature:

$$\text{efficiency} = \frac{T_{k1} - T_{k2}}{T_{k1}} \times 100\%$$

This efficiency is greater than that of most engines because real engines also have losses due to friction.

The second law of thermodynamics

The second law of thermodynamics can be stated thus: It is impossible to construct a heat engine that only absorbs heat from a heat source and performs an equal amount of work. In other words, no machine is ever 100 per cent efficient; some heat must be lost to the environment.

The second law also determines the order of physical phenomenon. Imagine viewing a film where a pool of water forms into an ice cube. Obviously, the film is running backward from the way in which it was filmed. An ice cube melts as it heats but never spontaneously cools to form an ice cube again; thus, this law indicates that certain events have a preferred direction of time, called the **arrow of time.** If two objects of different temperatures are placed in thermal contact, their final temperature will be between the original temperatures of the two objects. A second way to state the second law of thermodynamics is to say that heat cannot spontaneously pass from a colder to a hotter object.

Entropy

Entropy is the measure of how much energy or heat is unavailable for work. Imagine an isolated system with some hot objects and some cold objects. Work can be done as heat is transferred from the hot to the cooler objects; however, once this transfer has occurred, it is impossible to extract additional work from them alone. Energy is always conserved, but when all objects have the same temperature, the energy is no longer available for conversion into work.

The change in entropy of a system (ΔS) is defined mathematically as

$$\Delta S = \frac{\Delta Q}{T}$$

The equation states the following: The change in entropy of a system is equal to the heat flowing into the system divided by the temperature (in degrees Kelvin).

The entropy of the universe increases or remains constant in all natural processes. It is possible to find a system for which entropy decreases, but only due to a net increase in a related system. For example, the originally hotter objects and cooler objects reaching thermal equilibrium in an isolated system may be separated, and some of them put in a refrigerator. The objects would again have different temperatures after a period of time, but now the system of the refrigerator would have to be included in the analysis of the complete system. No net decrease in entropy of all the related systems occurs. This is yet another way of stating the second law of thermodynamics.

The concept of entropy has far-reaching implications that tie the order of our universe to probability and statistics. Imagine a new deck of cards in order by suits, with each suit in numerical order. As the deck is shuffled, no one would expect the original order to return. There is a probability that the randomized order of the shuffled deck would return to the original format, but it is exceedingly small. An ice cube melts, and the molecules in the liquid form have less order than in the frozen form. An infinitesimally small probability exists that all of the slower moving molecules will aggregate in one space so that the ice cube will reform from the pool of water. The entropy and disorder of the universe increase as hot bodies cool and cold bodies warm. Eventually, the entire universe will be at the same temperature, so the energy will be no longer usable.

Chapter Checkout

1. A glass bottle ($\alpha = 9 \times 10^{-6}/\,°C$) with a steel cap ($\alpha = 1.1 \times 10^{-5}/\,°C$) is placed into a freezer (–20° C) from room temperature (22°C). The rim of the bottle has radius 5 cm, and there is a 2 mm gap between the cap and the bottle initially. What is the change in the radii of each and in the size of the gap?
2. Helium from a tank with volume 1 m^3 and pressure 2400 Pa is used to fill a balloon to a volume of .001 m^3. What is the pressure in the balloon?
3. A 100 g block of ice is melted. What is the change in entropy during the melting process alone?
4. Two hot metal blocks of different composition, mass, and temperature are brought into thermal contact in complete isolation. Work out an expression for their final equilibrium temperature, given their masses, initial temperatures, and specific heat capacities.
5. What is the average speed of a nitrogen gas molecule at each of the following? (Note: Nitrogen gas has a molecular weight of 28 g/mole.)
   - **a.** 0°C
   - **b.** 22°C (room temperature)

Answers: 1. bottle rim: 1.9×10^{-3}cm; cap: 2.4×10^{-3}cm; gap: 5×10^{-4} cm **2.** 2.4×10^{6} Pa **3.** 122.3 J/K **4.** $T_f = \frac{c_1 m_1 T_1 - c_2 m_2 T_2}{c_1 m_1 - c_2 m_2}$ **5. a.** 402 m/s **b.** 512 m/s

Chapter 4

ELECTRICITY AND MAGNETISM

CHAPTER CHECK-IN

- ❑ Learning about electrostatics, including Coulomb's law and Gauss's law.
- ❑ Understanding the elements of dc and ac circuits: capacitors, resistors, and inductors.
- ❑ Calculating magnetic fields due to current sources, using Ampere's law.
- ❑ Applying Faraday's law and Lenz's law to understand electromagnetic induction.

The ancient Greeks observed that a small piece of amber, when rubbed, would pick up pieces of straw. Early peoples collected lodestone, a naturally occurring magnetic material, for its amazing ability to attract metals. Not until the nineteenth century did scientists establish that electricity and magnetism are related phenomena.

Now it is recognized that electromagnetic forces underlie most of our everyday experiences. Your eyes detect electromagnetic radiation. Electromagnetic repulsion ultimately prevents you from walking through a wall or falling through the floor. Each cell in your body acts as a small circuit, and your nervous system relies upon electrical impulses. And anyone who has been lost in the woods recognizes the usefulness of the Earth's magnetic field.

Although the formalism in this chapter is abstract, keep in mind that electric and magnetic phenomena are pervasive in the natural world. And, of course, understanding electromagnetism is critical to understanding most modern technology.

Electrostatics

Electrostatics, as the name implies, is the study of stationary electric charges. A rod of plastic rubbed with fur or a rod of glass rubbed with silk will attract small pieces of paper and is said to be **electrically charged.** The charge on plastic rubbed with fur is defined as **negative,** and the charge on glass rubbed with silk is defined as **positive.**

Electric charge

Electrically charged objects have several important characteristics:

- Like charges repel one another; that is, positive repels positive and negative repels negative.
- Unlike charges attract each another; that is, positive attracts negative.
- Charge is conserved. A neutral object has no net charge. If the plastic rod and fur are initially neutral, when the rod becomes charged by the fur, a negative charge is transferred from the fur to the rod. The net negative charge on the rod is equal to the net positive charge on the fur.

A **conductor** is a material through which electric charges can easily flow. An **insulator** is a material through which electric charges do not move easily, if at all. An **electroscope** is a simple device used to indicate the existence of charge. As shown in Figure 4-1, the electroscope consists of a conducting knob and attached lightweight conducting leaves—commonly made of gold foil or aluminum foil. When a charged object touches the knob, the like charges repel and force the leaves apart. The electroscope will indicate the presence of charge but does not directly indicate whether the charge is positive or negative.

Figure 4-1 An electroscope reports the presence of charge.

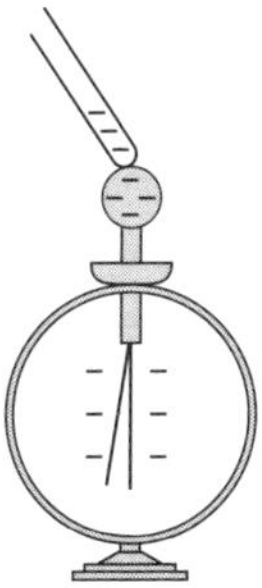

A large charge near a neutral electroscope can make the leaves move apart. The electroscope is made of conducting material, so the positive charges are attracted to the knob by the nearby (but not touching) negatively charged rod. The leaves are left with a negative charge and therefore deflect. When the negative rod is removed, the leaves will fall.

Now, consider touching the electroscope knob with a finger while the charged rod is nearby. The electrons will be repulsed and flow out of the electroscope through the hand. If the hand is removed *while the charged rod is still close*, the electroscope will retain a charge. This method of charging is called **charging by induction** (see Figure 4-2).

Figure 4-2 Charging an electroscope by induction.

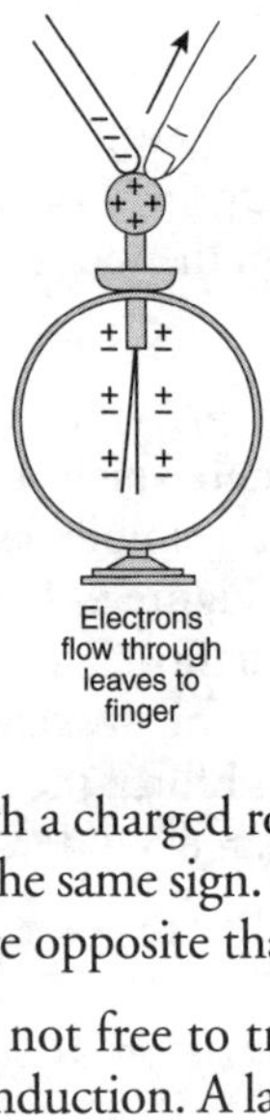

When an object is rubbed with a charged rod, the object shares the charge so that both have a charge of the same sign. In contrast, charging by induction gives an object the charge opposite that of the charged rod.

Even though the charges are not free to travel throughout the material, insulators can be charged by induction. A large charge nearby—not touching—will induce an opposite charge on the surface of the insulator. As shown in Figure 4-3, the negative and positive charges of the molecules are displaced slightly. This realignment of charges in the insulator produces an effective induced charge.

Figure 4-3 Induction of surface charge on an insulator by a nearby charged object.

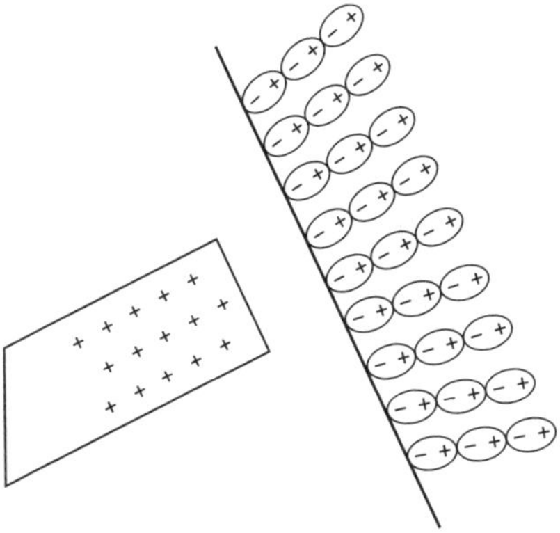

Coulomb's law

Coulomb's law gives the magnitude of the electrostatic force (F) between two charges:

$$F = \frac{kq_1 q_2}{r^2}$$

where q_1 and q_2 are the charges, r is the distance between them, and k is the proportionality constant. The SI unit for charge is the coulomb. If the charge is in coulombs and the separation in metres, the following approximate value for k will give the force in newtons: $k = 9.0 \times 10^9 \mathrm{N \cdot m^2/C^2}$. The direction of the electrostatic force depends upon the signs of the charges. Like charges repel, and unlike charges attract.

Coulomb's law can also be expressed in terms of another constant (ε_o), known as the **permittivity of free space:**

$$\varepsilon_o = \frac{1}{4\pi k} = 8.85 \times 10^{-12} \frac{\mathrm{C^2}}{\mathrm{N \cdot m^2}}$$

When the permittivity constant is used, Coulomb's law is

$$F = \frac{1}{4\pi\varepsilon_o} \frac{q_1 q_2}{r^2}$$

The most fundamental electric charge is the charge of one proton or one electron. This value *(e)* is $e = 1.602 \times 10^{-19}$ coulombs. It takes about 6.24×10^{18} excess electrons to equal the charge of one coulomb; thus, it is a very large static charge.

For a system of charges, the forces between each set of charges must be found; then, the net force on a given charge is the vector sum of these forces. The following problem illustrates this procedure.

Example 1: Consider equal charges of *Q* whose value in coulombs is not known. The force between two of these charges at distance *X* is *F*. In Figure 4-4, three charges (3*Q*) are placed at point *A*, which is a distance *X* from point *B*. One charge *(Q)* is placed at point *B*, which is *X*/2 distance from point *C*, which has one charge. What is the net force on the charge at point *B*?

Figure 4-4 Arrangement of point charges for the example.

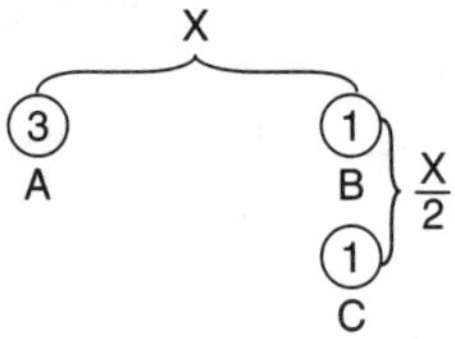

Solution: This problem can be solved through proportional reasoning. The force of 3*Q* on the one charge at *B* will be 3*F*. Because the single charge is one-half *X* from *B*, the force will be four times greater than at a distance *X*, that is, 4*F*. The forces of 3*F* and 4*F* are at right angles, and therefore, the resultant force is 5*F*, or

$$F_{net}\sqrt{(3F)^2 + (4F)^2} = 5F$$

The direction is found from the tangent: $\theta^{-1} = \tan 4/3 = 53°$.

Electric fields and lines of force

When a small positive test charge is brought near a large positive charge, it experiences a force directed away from the large charge. If the test charge is far from the large charge, the electrostatic force given by Coulomb's law is smaller than when it is near. This data of direction and magnitude of an electrostatic force, due to a fixed charge or set of fixed charges, constitutes an electrostatic field. The **electric field** is defined as the force per unit charge exerted on a small positive test charge (q_o) placed at that point. Mathematically,

$$E = \frac{F}{q_o}$$

Note that both the force and electric field are vector quantities. The test charge is required to be small so that the field of the test charge does not affect the field of the set charges being examined. The SI unit for electric field is newtons per coulomb (N/C).

Figure 4-5 is a pictorial representation of the electric fields surrounding a positive charge and a negative charge. These lines are called **field lines** or **lines of force**.

Figure 4-5 Electric field lines of (a) positive and (b) negative point charges.

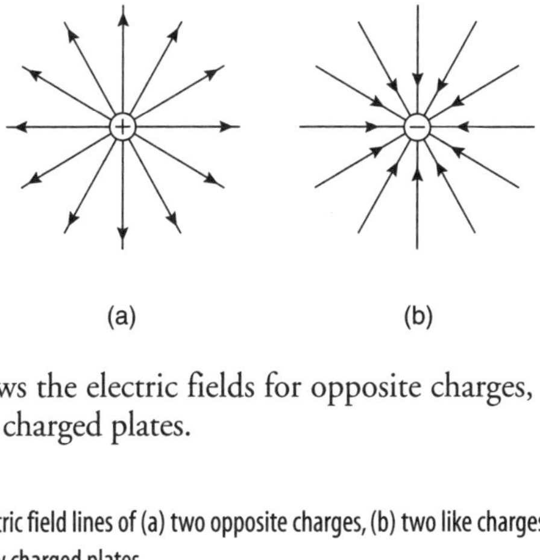

Figure 4-6 shows the electric fields for opposite charges, similar charges, and oppositely charged plates.

Figure 4-6 Electric field lines of (a) two opposite charges, (b) two like charges, and (c) two oppositely charged plates.

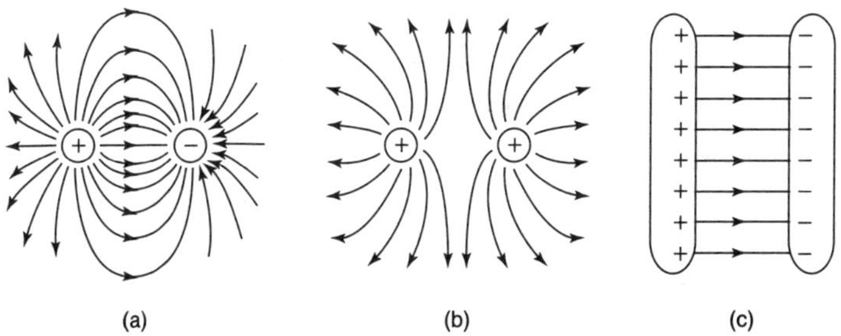

The rules for drawing electric field lines for any static configuration of charges are

- The lines begin on positive charges and terminate on negative charges.
- The number of lines drawn emerging from or terminating on a charge is proportional to the magnitude of the charge.
- No two field lines ever cross in a charge-free region. (Because the tangent to the field line represents the direction of the resultant force, only one line can be at every point.)
- The line approaches the conducting surface perpendicularly.

Electric flux

Electric flux is defined as the number of field lines that pass through a given surface. In Figure 4-7, lines of electric flux emerging from a point charge pass through an imaginary spherical surface with the charge at its centre.

Figure 4-7 Electric flux due to a point charge $+q$.

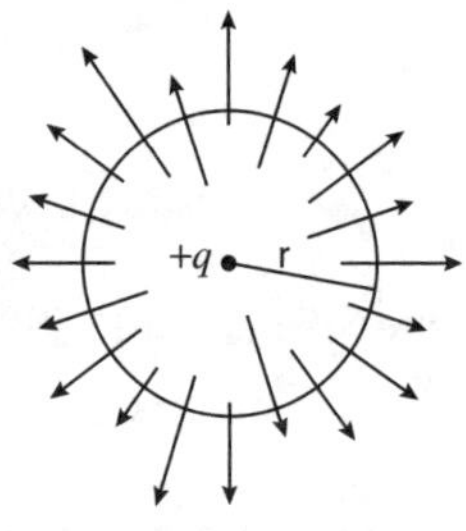

This definition can be expressed as follows: $\Phi = \sum \mathbf{E} \cdot \mathbf{A}$, where Φ (the Greek letter phi) is the electric flux, **E** is the electric field, and **A** is area perpendicular to the field lines. Electric flux is measured in $\mathrm{N \cdot m^2/C^2}$ and is a scalar quantity. If the surface under consideration is not perpendicular to the field lines, then the expression is $\Phi = \sum EA \cos\theta$.

In general terms, flux is the closed integral of the dot product of the electric field vector and the vector $\Delta\mathbf{A}$. The direction of $\Delta\mathbf{A}$ is the outward drawn normal to the imaginary surface. Mathematically, $\Phi = \oint \mathbf{E} \cdot d\mathbf{A}$. The accepted convention is that flux lines are positive if leaving a surface and negative if entering a surface.

Gauss's law

Gauss's law provides a method to calculate any electric field; however, its only practical use is for fields of highly symmetric distributions of fixed charges. The law states that the net electric flux through any real or imaginary closed surface is equal to the net electric charge enclosed within that surface divided by ε_o. As a result, if no charge exists with a given closed surface, then there are as many flux lines entering the surface as there are leaving it. The imaginary surface necessary to apply Gauss's law is called the **gaussian surface**. Algebraically,

$$\sum EA \cos\theta = \frac{Q}{\varepsilon_o}$$

or in integral form,

$$\int \mathbf{E} \cdot d\mathbf{A} = \frac{Q}{\varepsilon_o}$$

where θ is the angle between the direction of E and the outward direction of normal to the surface and ε_o is the permittivity constant.

Consider the calculation of the electric field due to a point charge. Figure 4-7 shows the point charge, the direction of its field, and a gaussian surface. Because the electric field is perpendicular to the gaussian surface and directed outward, θ is 90 degrees, and $\cos\theta = 1$. Gauss's law is

$$EA = \frac{q}{\varepsilon_o}$$

Substitute in the area of a sphere, and the left side reduces to

$$E4\pi r^2 = \frac{q}{\varepsilon_o}$$

or

$$E = \frac{q}{4\pi r^2 \varepsilon_o}$$

which is the same expression obtained from Coulomb's law and the definition of electric field in terms of force.

The derivation of the expression for the field due to a thin conducting shell of charge follows. Figure 4-8 shows the electric fields for (a) a shell of radius (R), (b) the gaussian surface for outside the shell, and (c) the gaussian surface for inside the shell (c) of radius (r).

Figure 4-8 (a) A charged spherical shell of radius R. (b) A gaussian sphere with radius $r > R$. (c) A gaussian sphere with radius $r < R$.

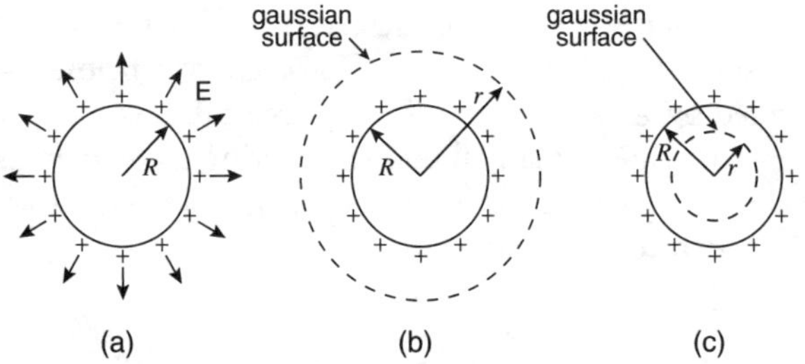

When outside the shell of charge, as in Figure 4-8(a), the left side of Gauss's equation reduces to the following expression for the same reasons given for a point charge:

$$E4\pi r^2 = \frac{q}{\varepsilon_o}$$

Therefore,

$$E = \frac{q}{4\pi r^2 \varepsilon_o}$$

Thus, the electric field outside a sphere of charge is the same as if the same amount of charge were concentrated in a point located at the centre of the sphere.

The gaussian surface inside the sphere encloses no charge, and therefore, there is no electric field inside the uniformly charged spherical shell. The same proof holds within a solid conductor because all the charge of the conductor resides on the surface. Because the electric field inside even an irregularly shaped conductor is zero, the charge will not be evenly distributed over an irregular shape. The charge will tend to accumulate on protruding points on the outside of the conductor.

Potential difference and equipotential surfaces

In the preceding examples, the charge distributions were spherical, and so the gaussian surface was a sphere. When finding the electric field of either a sheet of charge or a line of charge, a cylinder is the correct gaussian surface to use.

Example 2: Find the electric field for a nonconducting infinite sheet of charge. The electric field is directed outward from the sheet on both sides. The charge per unit area is σ (the Greek letter sigma). See Figure 4-9 for the electric field and gaussian surface.

Figure 4-9 A gaussian cylinder must extend above and below an infinite sheet of charge.

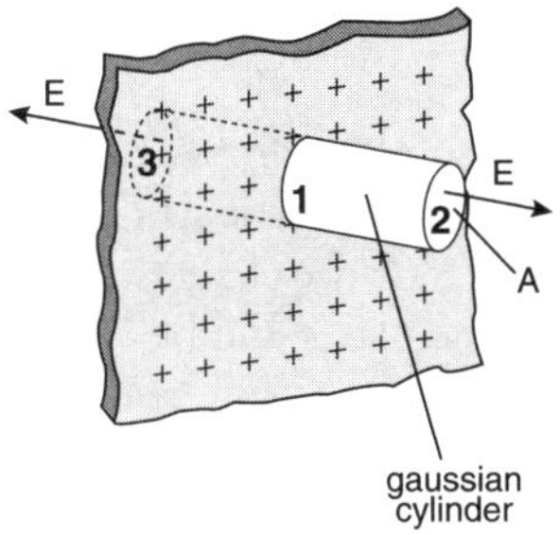

The area of the closed cylindrical gaussian surface will be the sum of the areas of the left end, the right end, and the wall; therefore, $\sum EA\cos\theta = (EA\cos\theta)_{left\ end} + (EA\cos\theta)_{right\ end} + (EA\cos\theta)_{wall}$. The electric field is parallel to the wall, which is at right angles to the outward normal of the wall area; thus, the last term on the right is zero. At each end, E is in the same direction as the outward normal, so $(EA\cos\theta)_{left\ end} + (EA\cos\theta)_{right\ end} = 2EA$, where A is the area of the end of the gaussian cylinder. The total charge inside the gaussian surface is the product of the charge per unit area and the area; so

$$2EA = \frac{\sigma A}{\varepsilon_o}$$

and

$$E = \frac{\sigma}{2\varepsilon_o}$$

Note that the magnitude of the electric field does not depend upon the distance from the plate. The electric field is uniform. In the practical case of finite plates of charge, the electric field is uniform relatively close to the charged plate.

The resultant electric field of two parallel plates is double that of one sheet with the same charge:

$$E = \frac{\sigma}{\varepsilon_o}$$

or

$$E = \frac{q}{\varepsilon_o A}$$

where q is the charge on each plate, and A is the area of each plate. If the plates have opposite charges, the electric field will exist between the plates and be zero outside the plates. If the charges are of equal sign, the electric field will be zero between the plates and be expressed by the above equation outside the plates. These results can be derived by Gauss's law.

Electrostatic potential and equipotential surfaces

Imagine moving a small test charge (q') from point A to point B in the uniform field between parallel plates. The work done in transferring the charge equals the product of the force on the test charge and the parallel component of displacement, using the same definition of work given in the section on mechanics (see Chapter 1). This work can also be expressed in terms of **E** from the definition of electric field as the ratio of force to charge: $W = \mathbf{F} \cdot \mathbf{d}$, $\mathbf{E} = \mathbf{F}/q$, and $W = q'\mathbf{E} \cdot \mathbf{d}$. See Figure 4-10.

Figure 4-10 Work is done when q' moves from position A to B in an electric field **E**.

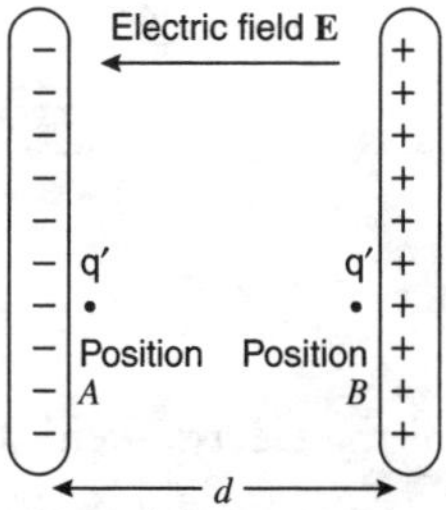

Work is change in potential energy: $U_B - U_A = q'Ed$.

In general, the **electrostatic potential difference**, sometimes called the **electric potential difference**, is defined as the energy change per unit positive charge, or $V_B - V_A = (U_B - U_A)/q'$. For certain configurations of electric field, it may be necessary to use the integral definition of electrostatic potential:

$$V = -\int_A^B \mathbf{E} \cdot ds$$

where a test charge moves over a line integral from point A to point B along path **s** in an electric field (**E**).

For the special case of parallel plates:

$$V = Ed = \frac{qd}{\varepsilon_o A}$$

where V is the potential difference between the plates, measured in units of volts (V):

$$1 \text{ volt} = 1 \frac{\text{joule}}{\text{coulomb}}$$

The electric potential due to a point charge *(q)* at a distance *(r)* from the point charge is

$$V = k\frac{q}{r}$$

The following problem illustrates the calculations of electric field and potential due to point charges.

Example 3: Given two charges of $+3Q$ and $-Q$, a distance X apart, find the following: (1) At what point(s) along the line is the electric field zero? (2) At what point(s) is the electric potential zero? (See Figure 4-11.)

Figure 4-11 The arrangement of two point charges for the example.

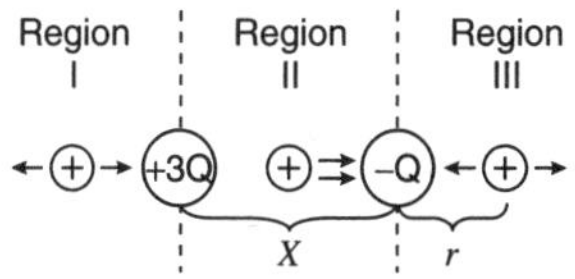

The first task is to find the region(s) where the electric field is zero. The electric field is a vector, and its direction can be located by a test charge. Figure 4-11 is divided into three regions. Between the opposite charges, the direction of the force on the test charge will be in the same direction from each charge; therefore, it is impossible to have a zero electric field in Region II. Even though the forces on the test charge from the two charges are in opposite directions in Region I, the force, and therefore the electric field, can never be zero in this region because the test charge is always closer to the largest given charge. Therefore, Region III is the only place where E can be zero. Select an arbitrary point *(r)* to the right of $-Q$ and set the two electric fields equal. Because the fields are in opposite directions, the vector sum at this point will equal zero.

$$E_{3Q} = E_Q$$

$$\frac{k(3Q)}{(x+r)^2} = \frac{k(Q)}{r^2}$$

$$\frac{r}{X+r} = \sqrt{\frac{1}{3}}$$

If X is given, solve for r.

Potential is not a vector, so the potential is zero wherever the following equation holds:

$$0 = \frac{k(+3Q)}{r_1} + \frac{k(-Q)}{r_2}$$

where r_1 is the distance from the test point to $+3Q$ and r_2 is the distance to $-Q$.

This example illustrates the difference in methods of analysis in finding the vector quantity (**E**) and the scalar quantity *(V)*. Note that if the charges were either both positive or both negative, it would be possible to find a point with zero electric field between the charges, but the potential would never be zero.

The **electrical potential energy** of a pair of point charges separated by a distance r is

$$P.E. = k\frac{q_1 q_2}{r}$$

Equipotential surfaces are surfaces where no work is required to move a charge from one point to another. The equipotential surfaces are always perpendicular to the electric field lines. **Equipotential lines** are two-dimensional representations of the intersection of the surface with the plane of the diagram. In Figure 4-12, equipotential lines are shown for (a) a uniform field, (b) a point charge, and (c) two opposite charges.

Figure 4-12 Equipotential lines for (a) a uniform electric field, (b) a point charge, and (c) a dipole.

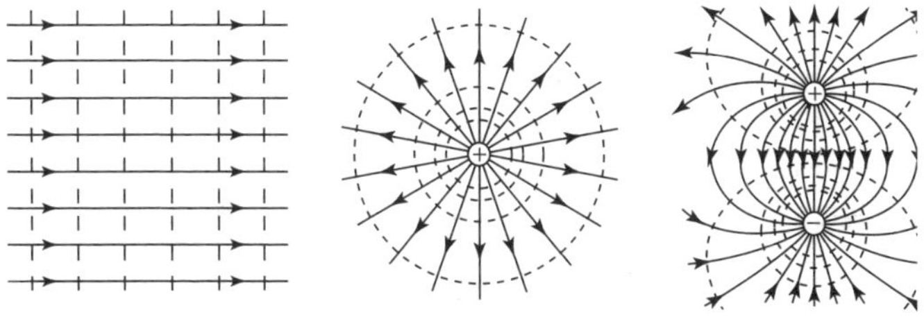

(a) (b) (c)

Capacitors

A **capacitor** is an electrical device for storing charge. In general, capacitors are made from two or more plates of conducting material separated by a layer or layers of insulators. The capacitor can store energy to be given to a circuit when needed.

Capacitance

The **capacitance** *(C)* is defined as the ratio of the stored charge *(Q)* to the potential difference *(V)* between the conductors:

$$C = \frac{Q}{V}$$

Capacitance is measured in farads *(F)* and 1 farad = 1 $\frac{\text{coulomb}}{\text{volt}}$.

Parallel plate capacitor

In its simplest form, the capacitor is a set of oppositely charged parallel plates separated by a distance *(d)*. From the equation for the potential difference of parallel plates and the definition of capacitance, the capacitance for parallel plates is

$$C = \frac{\varepsilon_o A}{d}$$

Strictly speaking, this equation is valid only when there is a vacuum between the plates.

When a nonconducting material is placed between the capacitor plates, more charge can be stored because of the induced charge on the surface of the electrical insulator. The ratio of the capacitance with the insulator to the vacuum capacitance is called the **dielectric constant** (κ, the Greek letter kappa). The values for the dielectric constants can be found in tables of properties of materials. The equation for the **parallel plate capacitor** with a dielectric that fills the space between the plates is

$$C = \frac{\kappa \varepsilon_o A}{d}$$

The energy stored in a capacitor can be found by any of the following three equations, which are each in terms of different variables:

$$\text{Energy stored} = \frac{1}{2}QV = \frac{1}{2}CV^2 = \frac{Q^2}{2C}$$

Parallel and series capacitors

Capacitors can be connected either in parallel or in series. Two capacitors are in **parallel** if the negative plates are connected and the positive plates are connected, as shown in Figure 4-13.

An equation can be derived for the capacitance of one capacitor that would have the equivalent capacitance of these two capacitors. The total charge stored on the two capacitors is $Q = Q_1 + Q_2$. The voltage across each capacitor is the same and is equal to the voltage of the battery *(V)*; therefore, $Q_1 = C_1 V$ and $Q_2 = C_2 V$, or for the equivalent capacitor, $Q = C_{eq} V$. Substituting into the equation for total charge yields $C_{eq} V = C_1 V + C_2 V$, or $C_{eq} = C_1 + C_2$. This result can be generalized to state that the equivalent capacitance for a set of capacitances in parallel is simply the sum of the individual capacitances.

Figure 4-13 Two capacitors connected in parallel. The drawing (a) is equivalent to the schematic (b).

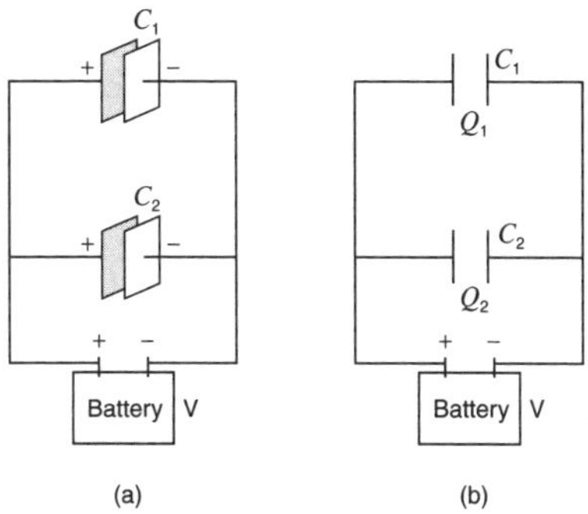

Capacitors are connected in **series** if the positive plate of one is connected to the negative plate, as shown in Figure 4-14.

Figure 4-14 Two capacitors connected in series. The drawing (a) is equivalent to the schematic (b).

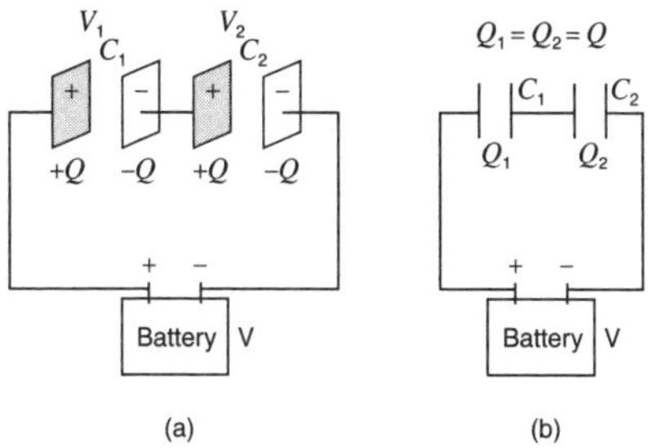

In series combinations, all the capacitors have the same charge. The potential differences across the capacitors add to equal the potential difference between the terminals of the battery; therefore,

$$V = V_1 + V_2,\ V_1 = \frac{Q}{C_1},\ \text{and}\ V_2 = \frac{Q}{C_2}$$

Substituting these equations into the equation for potential difference gives

$$\frac{Q}{C_{eq}} = \frac{Q}{C_1} + \frac{Q}{C_2}$$

Canceling the charge gives the following expression for the equivalent capacitance for series combinations:

$$\frac{1}{C_{eq}} = \frac{1}{C_1} + \frac{1}{C_2}$$

Note: A common mistake in calculating series capacitance is to forget to take the reciprocal to find the equivalent capacitance after adding the reciprocals of the individual capacitors.

Current and Resistance

Electric circuits with charges in motion are commonplace in our technological society. Current, resistance, and electromotive force are concepts necessary to describe these circuits.

Current

Current *(I)* is the amount of charge per time that passes through an area perpendicular to the flow:

$$I = \frac{\Delta q}{\Delta t}$$

Current is measured in SI units of amperes *(A),* and

$$1 \text{ ampere} = 1 \, \frac{\text{coulomb}}{\text{second}}$$

This definition for current can be applied to charges moving in a wire, in an electrolytic cell, or even in ionized gases.

In visualizing charge flowing through a circuit, it is *not* accurate to imagine the electrons moving very rapidly around the circuit. The average velocity, or **drift velocity** (v_d), of individual charges is low; the conduction electrons in a copper wire move on the order of 10^{-4} m/s. The formula is

$$v_d = \frac{I}{nqA}$$

where q is the charge on an electron, A is the cross-sectional area of the wire, and n is the number of conduction electrons per cubic metre. At this rate, the time to travel 10 cm is about 11 minutes. It is obvious from experience that it does not take this long for a bulb to glow after the switch is closed. When the circuit is completed, the entire charge distribution responds almost immediately to the electric field and is set in motion almost simultaneously, even though individual charges move slowly.

The battery provides a voltage *(V)* between its terminals. The electric field set up in a wire connected to the battery terminals causes the current to flow, which occurs when the current has a complete conducting path from one terminal of the batter to the other—called a **circuit**. By convention, the direction of current in the external circuit (not in the battery) is the direction of motion of positive charges. In metals, the electrons are the moving charges, so the definition of the direction of current is opposite the actual flow of the negative charges in a wire. (***Note:*** Electric fields are not found in conductors with static charges as shown by Gauss's law, but electric fields can exist in a conductor when charges are in motion.)

The potential difference between the terminals of the battery when no current is present is called the **electromotive force** (emf). The historical term *emf* is a misnomer because it is measured in volts, not force units, but the terminology is still commonly used.

Resistance and resistivity

Experimentally, it was found that current is proportional to voltage for conductors. The proportionality constant is the **resistance** in the circuit. This relationship is called **Ohm's law**: $V = IR$. Resistance is measured in ohms (Ω): an ohm is equal to 1 volt/1 ampere.

The resistance of a conductor depends upon its length *(l)*, its cross-sectional area *(A)*, and its resistivity (ρ). The **resistivity** for a specific conductor can be found in a table of properties of materials. The unit of resistivity is the ohm-metre. Resistance to current in a conductor arises because the flow of moving charges is impeded by the material of the wire. It is intuitive that the resistance should increase with the length of the wire, be inversely proportional to the cross-sectional area (less resistance for a larger area), and depend upon the wire substance. The relationship between resistance and resistivity is

$$R = \frac{l}{A}\rho$$

Note: A *resistor* is a specific electronic component whose only function is to resist current. A *resistance* is generated by anything impeding current, for example, a light bulb or heating element.

Electrical power and energy

Figure 4-15 shows a simple circuit of a battery with wires connecting it to a bulb. The filament in the bulb is a resistance shown in the circuit as R beside the symbol for a resistance —ww— . The symbol for the voltage of the battery is ε. Assume that the resistance in the connecting wires is negligible so that the light bulb is effectively the only resistance in the circuit. A constant potential difference is supplied by the battery—say, for example, 6 volts. When the current passes through the light bulb, charges move from a higher potential to a lower, with a difference of 6 volts. Energy is being converted into light and heat by the bulb filament.

Figure 4-15 A simple circuit with a light bulb represented by the resistor *R*.

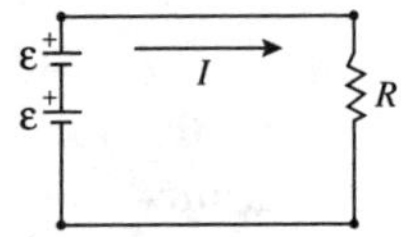

The rate of energy expenditure is **power,** given by any of the three expressions:

$$P = IV = \frac{V^2}{R} = I^2R$$

Power is measured in units of watts (W):

$$1 \text{ watt} = 1 \text{ joule/second} = 1 \text{ ampere-volt}$$

Direct Current Circuits

The simple circuit necessary to light a bulb with a battery was discussed in the previous paragraphs. The battery provides **direct current,** a current flowing in only one direction. This section is concerned with the analysis of simple direct current circuits of two types: (1) those with combinations of resistor elements and (2) those with batteries in different branches of a multiple-loop circuit.

Series and parallel resistors

Resistance, at least to some degree, exists in all electrical elements. The resistors might be light bulbs, heating elements, or components specifically manufactured for their resistance. It is assumed that the resistance in the connecting wires is negligible.

The series connection of two resistors (R_1 and R_2) is shown in Figure 4-16. What is the equivalent resistor for this combination?

Because there is only one pathway for the charges, the current is the same at any point in the circuit, that is, $I = I_1 = I_2$. The potential difference supplied by the battery equals the potential drop over R_1 and the potential drop over R_2. Thus,

$$V = V_1 + V_2$$

from Ohm's law, $$V = IR_1 + IR_2$$

and $$V = I(R_1 + R_2)$$

therefore, $$R_{eq} = R_1 + R_2$$

Figure 4-16 Two resistors connected in series. The drawing (a) is equivalent to the schematic (b).

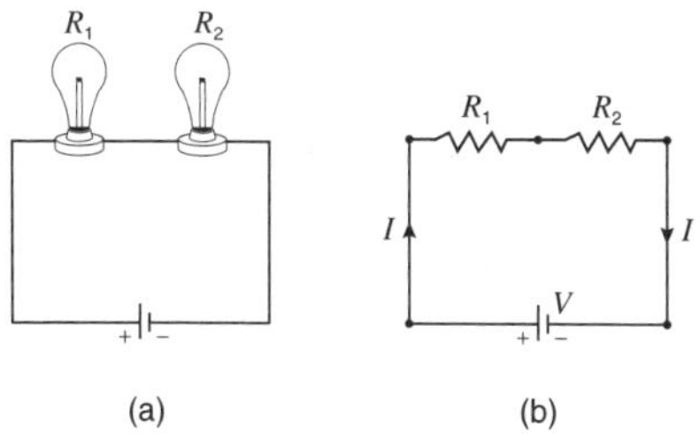

When resistors are in series, the equivalent resistance is the sum of the individual resistances. Compare this result with adding capacitors in series. For series resistors, the current is the same; while for series capacitors, the charge is the same. (Note that the equivalent resistance is a simple sum, but the equivalent capacitance is given by a reciprocal expression.)

The parallel connection for two resistors (R_1 and R_2) is shown in Figure 4-17. What is the equivalent resistance for this combination?

At point *a* for the circuit diagram—see Figure 4-17 (b)—the current branches so that part of the total current in the circuit goes through the upper branch and part through the lower branch. The potential drop of the current is the same regardless of which path is taken; therefore, the voltage difference is the same over either resistor ($V_{batt} = V_1 = V_2$). The currents sum to the total current:

$$I = I_1 + I_2$$

from Ohm's law, $I = \frac{V}{R_1} + \frac{V}{R_2}$, and $I = V\left(\frac{1}{R_1} + \frac{1}{R_2}\right)$

therefore, $$\frac{1}{R_{eq}} = \frac{1}{R_1} + \frac{1}{R_2}$$

Figure 4-17 Two resistors connected in parallel. The drawing (a) is equivalent to the schematic (b).

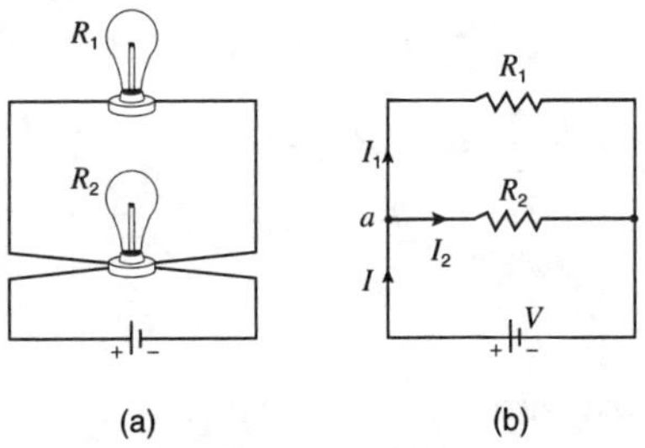

Thus, the reciprocal of the equivalent resistance is equal to the sum of the reciprocals of the individual resistors in the parallel combination. Compare this result with adding capacitors in parallel. For parallel resistors, the voltages across the resistors are equal, and the same is true for parallel capacitors. (Note that the equivalent resistance is a reciprocal expression, but the equivalent capacitance for parallel combination is a simple sum.)

Kirchhoff's rules

If a circuit has several batteries in the branches of multiloop circuits, the analysis is greatly simplified by using **Kirchhoff's rules,** which are forms of conservation laws:

- The sum of the currents entering a junction must equal the sum of the currents leaving the junction. This rule, sometimes called the **junction rule,** is a statement of conservation of charge. Because charge neither builds up at any place in the circuit nor leaves the circuit, the charge entering a point must also leave that point.
- The algebraic sum of the drops in potential across each element around any loop must equal the algebraic sum of the emfs around any loop. This rule expresses conservation of energy. In other words, the charge moving around any loop must gain as much energy from batteries as it loses when going through resistors.

When applying Kirchhoff's rules, use consistent sign conventions. Refer to the directions selected for the currents in Figure 4-18. Fewer mistakes will be made if one direction is consistently used—for example, clockwise in all loops. If an incorrect direction for one current is selected initially, the solution for that current will be negative. Use the following sign conventions when applying the loop rule:

- If the resistor is traveled in the direction of the current, the change in potential is negative, and if traveled opposite to the selected direction of the current, it is positive.
- If a source of emf is traveled in the direction of the emf (from – to + between the terminals), then the change in potential is positive, and if traveled opposite to the direction of the emf, it is negative.

Check the equations for Figure 4-18.

Figure 4-18 A circuit illustrating the application of Kirchhoff's rules, and the resulting equations.

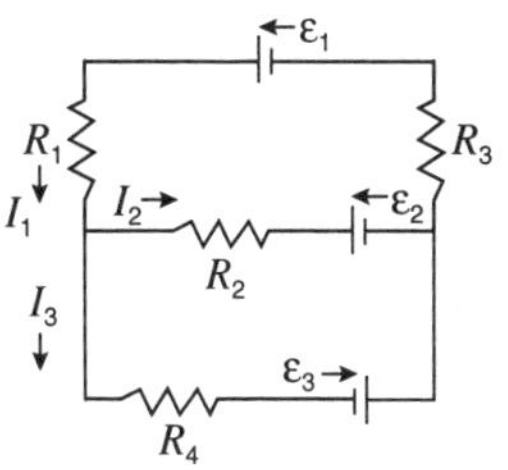

$$I_1 = I_2 + I_3$$

$$\text{[top loop]}\ I_1R_1 + I_2R_2 + I_1R_3 = \varepsilon_1 - \varepsilon_2$$

$$\text{[bottom loop]}\ I_3R_4 - I_2R_2 = \varepsilon_3 + \varepsilon_2$$

$$\text{[outside loop]}\ I_1R_1 + I_3R_4 + I_1R_3 = \varepsilon_1 + \varepsilon_3$$

Imagine that the values of the resistances and voltage were given for this problem. Then, it would be possible to write four different equations: the junction equation, the top loop, the bottom loop, and the outside loop. Only three currents exist, however, so only three equations are necessary. In this case, solve the set of equations that are the easiest to manipulate.

Electromagnetic Forces and Fields

The magnetic field of naturally occurring magnetite is too weak to be used in devices such as modern motors and generators; these magnetic fields must come from electric currents. Magnetic fields affect moving charges, and moving charges produce magnetic fields; therefore, the concepts of magnetism and electricity are closely intertwined.

Magnetic fields and lines of force

A bar magnet attracts iron objects to its ends, called **poles.** One end is the **north pole,** and the other is the **south pole.** If the bar is suspended so that it is free to move, the magnet will align itself so that its north pole points to the geographic north of the earth. The suspended bar magnet acts like a compass in the earth's magnetic field. If two bar magnets are brought close together, the like poles will repel each other, and the unlike poles attract each other. (***Note:*** By this definition, the magnetic pole under the earth's north geographical pole is the south pole of the earth's magnetic field.)

This magnetic attraction or repulsion can be explained as the effect of one magnet on the other, or it can be said that one magnet sets up a **magnetic field** in the region around it that affects the other magnet. The magnetic field at any point is a vector. The direction of the magnetic field (**B**) at a specified point is the direction that the north end of a compass needle points at that position. **Magnetic field lines,** analogous to electric field lines, describe the force on magnetic particles placed within the field. Iron filings will align to indicate the patterns of magnetic field lines.

Force on a moving charge

If a charge moves through a magnetic field at an angle, it will experience a force. The equation is given by $\mathbf{F} = q\mathbf{v} \times \mathbf{B}$ or $F = qvB \sin\theta$, where q is the charge, **B** is the magnetic field, **v** is the velocity, and θ is the angle between the directions of the magnetic field and the velocity; thus, using the definition of the cross product, the definition for the magnetic field is

$$B = \frac{F}{qv\sin\theta}$$

Magnetic field is expressed in SI units as a tesla (T), which is also called a weber per square metre:

$$\mathrm{T} = \frac{\mathrm{Wb}}{\mathrm{m}^2} = \frac{\mathrm{N}}{\mathrm{Cm/s}} = \frac{\mathrm{N}}{\mathrm{Am}}$$

The direction of **F** is found from the right-hand rule, shown in Figure 4-19.

To find the direction of the force on the charge, with a flat hand point your thumb in the direction of the velocity of the positive charge and your fingers in the direction of the magnetic field. The direction of the force is out of the palm of your hand. (If the moving charge is negative, point your thumb opposite to its direction of motion.) Mathematically, this force is the cross product of the velocity vector and the magnetic field vector.

Figure 4-19 Using the right-hand rule to find the direction of magnetic force on a moving charge.

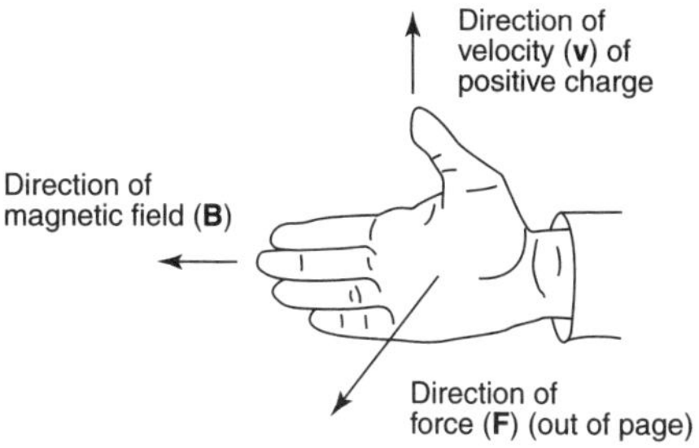

If the velocity of the charged particle is perpendicular to the uniform magnetic field, the force will always be directed toward the centre of a circle of radius r, as shown in Figure 4-20. The x symbolizes a magnetic field into the plane of the paper—the tail of the arrow. (A dot symbolizes a vector out of the plane of the paper—the tip of the arrow.)

Figure 4-20 The force on a charge moving perpendicular to a magnetic field is toward the centre of a circle.

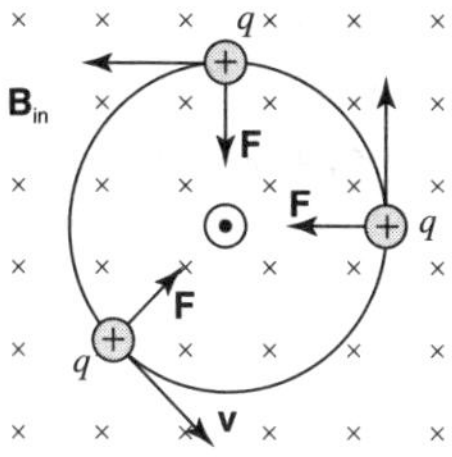

The magnetic force provides centripetal acceleration:

$$F = qvB = \frac{mv^2}{r}$$

or

$$r = \frac{mv}{qB}$$

The radius of the path is proportional to the mass of the charge. This equation underlies the operation of a **mass spectrometer,** which can separate equally ionized atoms of slightly different masses. The singly ionized atoms are given equal velocities, and because their charges are the same and they travel through the same **B**, they will travel in slightly different paths and can then be separated.

Force on a current-carrying conductor

Charges confined to wires can also experience a force in a magnetic field. A current (I) in a magnetic field (**B**) experiences a force (**F**) given by the equation $\mathbf{F} = I\mathbf{l} \times \mathbf{B}$ or $F = IlB \sin\theta$, where **l** is the length of the wire, represented by a vector pointing in the direction of the current. The direction of the force may be found by a right-hand rule similar to the one shown in Figure 4-19. In this case, point your thumb in the direction of the current—the direction of motion of positive charges. The current will experience no force if it is parallel to the magnetic field.

Torque on a current loop

A loop of current in a magnetic field can experience a torque if it is free to turn. Figure 4-21(a) depicts a square loop of wire in a magnetic field directed to the right. Imagine in Figure 4-21(b) that the axis of the wire is turned to an angle (θ) with the magnetic field and that the view is looking down on the top of the loop. The *x* in a circle depicts the current traveling into the page away from the viewer, and the dot in a circle depicts the current out of the page toward the viewer.

Figure 4-21 (a) Square current loop in a magnetic field **B**. (b) View from the top of the current loop. (c) If the loop is tilted with respect to **B**, a torque results.

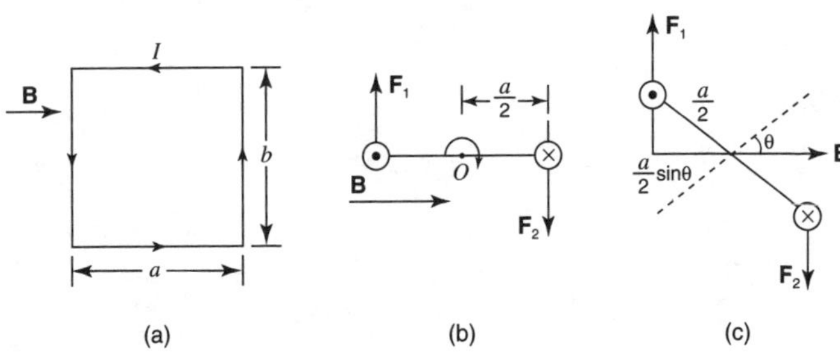

The right-hand rule gives the direction of the forces. If the loop is pivoted, these forces produce a torque, turning the loop. The magnitude of this torque is $\mathbf{t} = NI\mathbf{A} \times \mathbf{B}$, where N is the number of turns of the loop, $\mathbf{B}$ is the magnetic field, I is the current, and $\mathbf{A}$ is the area of the loop, represented by a vector perpendicular to the loop.

Galvanometers, ammeters, and voltmeters

The torque on a current loop in a magnetic field provides the basic principle of the **galvanometer,** a sensitive current-measuring device. A needle is affixed to a current coil—a set of loops. The torque gives a certain deflection of the needle, which is dependent upon the current, and the needle moves over a scale to allow a reading in amperes.

An **ammeter** is a current-measuring instrument constructed from a galvanometer movement in parallel with a resistor. Ammeters are manufactured to measure different ranges of current. A **voltmeter** is constructed from a galvanometer movement in series with a resistor. The voltmeter samples a small portion of the current, and the scale provides a reading of potential difference—volts—between two points in the circuit.

Magnetic field of a long, straight wire

A current-carrying wire generates a magnetic field of magnitude B in circles around the wire. The equation for the magnetic field at a distance r from the wire is

$$B = \frac{\mu_o I}{2\pi r}$$

where I is the current in the wire and μ_o (the Greek letter mu) is the proportionality constant. The constant, called the **permeability constant,** has the value

$$\mu_o = 4\pi \times 10^{-7}\, T\,\frac{m}{A}$$

The direction of the field is given by a second right-hand rule, shown in Figure 4-22.

Grasp the wire so that your thumb points in the direction of the current. Your fingers will curl around the wire in the direction of the magnetic field.

Figure 4-22 Using the second right-hand rule to determine the direction of the magnetic field resulting from a current.

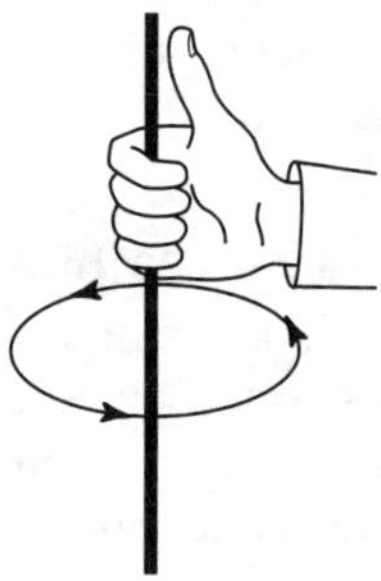

Ampere's law

Ampere's law allows the calculation of magnetic fields. Consider the circular path around the current shown in Figure 4-22. The path is divided into small elements of length (Δl). Note the component of **B** that is parallel to Δl and take the product of the two to be $B_{\parallel}\Delta l$. Ampere's law states that the sum of these products over the closed path equals the product of the current and μ_o;

$$\sum B_{\parallel}\Delta l = \mu_o I$$

Or in integral form,

$$\oint \mathbf{B} \cdot \mathrm{dl} = \mu_0 \mathrm{I}$$

Somewhat analogous to the way Gauss's law can be used to find the electric field for highly symmetric charge configurations, Ampere's law can be used to find the magnetic fields for current configurations of high symmetry. For example, Ampere's law can be used to derive the expression for the magnetic field generated by a long, straight wire:

$$\sum B_{\parallel}\Delta l = \sum \Delta l = B_{\parallel}(2\pi r) = \mu_0 I$$

$$B = \frac{\mu_o I}{2\pi r}$$

Magnetic fields of the loop, solenoid, and toroid

A current generates a magnetic field, and the field differs as the current is shaped into (a) a loop, (b) a solenoid (a long coil of wire), or (c) a toroid (a donut-shaped coil of wire). The equations for the magnitudes of these fields follow. The direction of the field in each case can be found by the second right-hand rule. Figure 4-23 illustrates the fields for these three different configurations.

(a) The field at the centre of a single loop is given by

$$B = \mu_0 \frac{I}{2r}$$

where r is the radius of the loop.

(b) The field due to a solenoid is given by $B = \mu_0 NI$, where N is the number of turns per unit length.

(c) The field due to a toroid is given by

$$B = \mu_0 \frac{NI}{2\pi R}$$

where R is the radius to the centre of the toroid.

Figure 4-23 Magnetic field resulting from (a) a current loop, (b) a solenoid, and (c) a toroid.

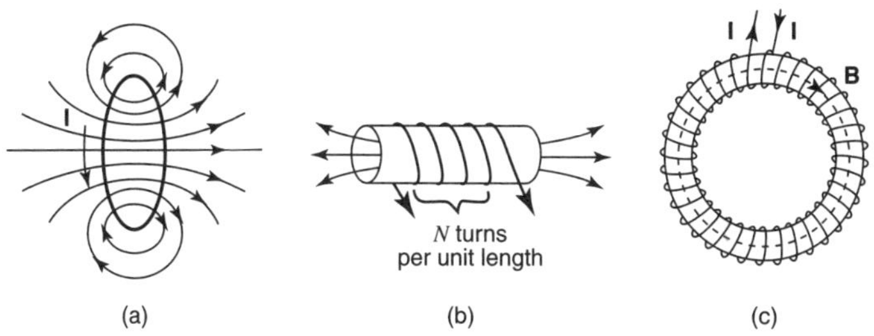

Electromagnetic Induction

The finding that electric current can produce magnetic fields led to the idea that magnetic fields could produce electric currents. The production of emfs and currents by the changing magnetic field through a conducting loop is called **induction.**

Magnetic flux

The magnetic field through a loop can be changed either by changing the magnitude of the field or by changing the area of the loop. To be able to quantitatively describe these changes, **magnetic flux** is defined as $\Phi = BA\cos\theta$, where θ is the angle between **B** and the direction perpendicular to the plane of the loop (along the axis of the loop).

Faraday's law

Changing the magnetic flux through a loop of wire induces a current. **Faraday's law** states that the emf induced in a wire is proportional to the rate of the flux through the loop. Mathematically,

$$\varepsilon = -N\frac{\Delta\Phi}{\Delta t}$$

where N is the number of loops, $\Delta\Phi$ is the change of flux in time, Δt. The minus sign indicates the polarity of the induced emf.

The preceding equation is easy to use when the flux is set up by an electromagnet. If the electromagnet is turned on or off, the induced emf is equal to the number of turns in the loop times the rate of change of flux. The flux might also be changed through a loop altering the size of the loop. Imagine a slide wire, as shown in Figure 4-24, where l is the length of the wire that moves in contact with the U-shaped wire. In this case, $\varepsilon = Blv$, where v is the velocity of the sliding length.

Figure 4-24 A slide wire circuit generates emf.

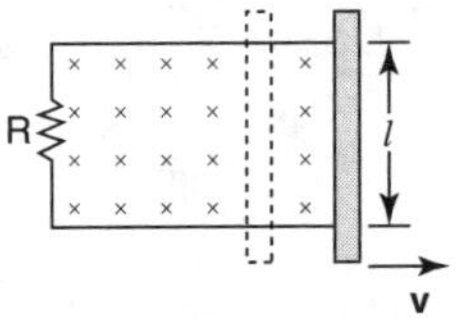

Note that this induced emf is indistinguishable from that of a battery and that the current is still just the rate of the motion of charges; therefore, Ohm's law and other relationships for currents in wires are still valid.

Lenz's law

The direction of the induced current can be found from Lenz's law, which states that the magnetic field generated by the induced emf produces a current whose magnetic field opposes the original change in flux through the wire loop. Again, consider Figure 4-24 and assume the slide is moving to the right. The *x* shapes indicate that **B** is into the page; thus, when the slide moves to the right, the field through the slide will get greater into the page. (The change in flux is the crucial quantity.) The magnetic field from the induced current will be directed out of the page because it will oppose the change in flux. Use the second-hand rule and place the curl of the fingers out of the page at the centre of the loop. The direction of the thumb indicates that the current will flow counterclockwise. (It is not correct to state that the current is to the right because it is to the left on the top of the loop.) Conversely, if the slide moves to the left, **B** will decrease through the loop. The change in flux will be out of the page, and the induced current will be clockwise. The same analysis is used if an electromagnet is turned on or off.

Lenz's law is also a conservation law. If the magnetic field generated by the induced current could be in the same direction as the original change in flux, the change would get larger and the induced current greater. This impossible condition would be a better energy exchange than a perpetual motion machine.

Generators and motors

Generators and motors are applications of electromagnetic induction. Figure 4-25 illustrates a simple electric generator.

The crank represents a mechanical method of turning the loop of wire in a magnetic field. The change in magnetic flux through the loop generates an induced current; thus, the **generator** converts mechanical energy into electrical energy. The operation of a motor is similar to that of a generator but in reverse. The motor has similar physical components except that the electric current supplied to the loop exerts a torque, which turns the loop. The **motor,** therefore, converts electrical energy into mechanical energy.

Figure 4-25 A simple electric generator.

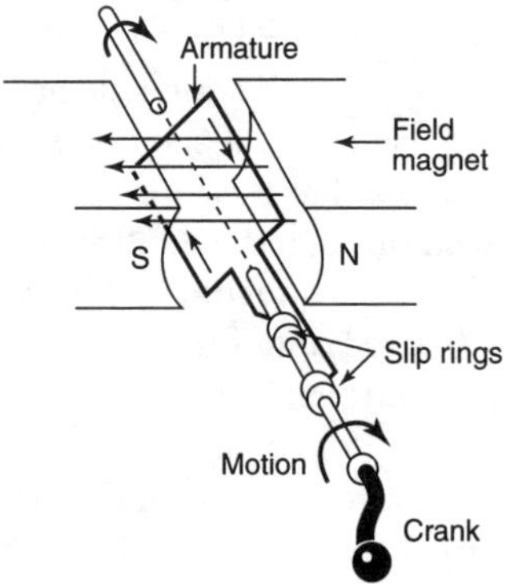

Mutual inductance and self-inductance

Mutual inductance occurs when two circuits are arranged so that the change in current in one causes an emf to be induced in the other.

Imagine a simple circuit of a switch, a coil, and a battery. When the switch is closed, the current through the coil sets up a magnetic field. As the current is increasing, the magnetic flux through the coil is also changing. This changing magnetic flux generates an emf opposing that of the battery. This effect occurs only while the current is either increasing to its steady state value immediately after the switch is closed or decreasing to zero when the switch is opened. This effect is called **self-inductance.** The proportional constant between the self-induced emf and the time rate of change of the current is called **inductance** *(L)* and is given by

$$\varepsilon = -L\frac{\Delta I}{\Delta t}$$

The SI unit for inductance is the henry, and 1 henry = 1(Vs/A).

Using Faraday's law, inductance can be expressed in terms of the change of flux and current:

$$L = \frac{N\Phi}{I}$$

where N is the number of turns of the coil.

Maxwell's equations and electromagnetic waves

Maxwell's equations summarize electromagnetic effects in four equations. The equations are too complex for this text, but the concepts embodied in them are important to consider. Maxwell explained that electric and magnetic waves can be generated by oscillating electric charges. These electromagnetic waves may be depicted as crossed electric and magnetic fields propagating through space perpendicular to the direction of motion and to each other, as illustrated in Figure 4-26.

Figure 4-26 An electromagnetic wave consists of perpendicular oscillating magnetic and electric fields.

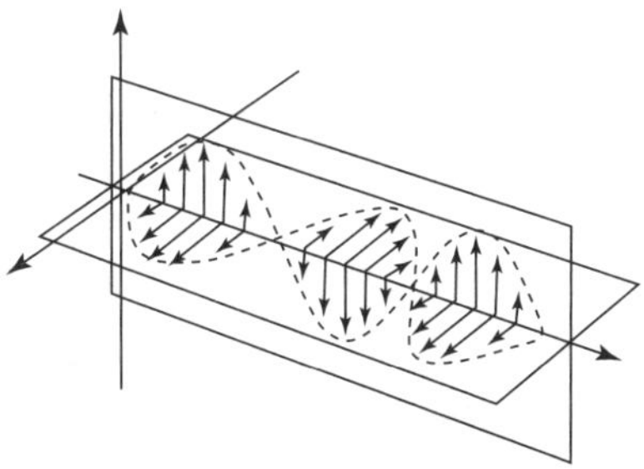

Alternating Current Circuits

Alternating currents and voltages are sinusoidal and vary with time. Alternating currents produce different responses in resistors, capacitors, and inductors than do the direct currents studied in previous sections.

Alternating currents and voltages

Figure 4-27 shows the plot of **alternating voltage** and **alternating current** as a function of time in a circuit that has only a resistor and a source of alternating current — an ac generator.

Figure 4-27 Current and voltage from an ac source through a simple resistor.

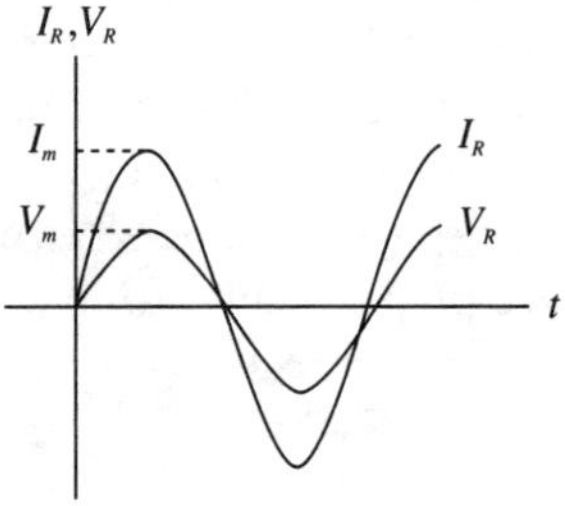

Because the voltage and current reach their maximum values at the same time, they are **in phase.** Ohm's law and the previous expressions for power are valid for this circuit if the **root mean square** (rms) of the voltage and the rms of the current, sometimes called the **effective value,** are used. These relationships are

$$I_{rms} = \frac{I_{max}}{\sqrt{2}} \text{ and } V_{rms} = \frac{V_{max}}{\sqrt{2}}$$

Ohm's law is expressed thus: $V_R = IR$, where V_R is the rms voltage across the resistor and I is the rms in the circuit.

Resistor-capacitor circuits

A circuit with a resistor, a capacitor, and an ac generator is called an **RC circuit.** Remember that a capacitor is basically a set of conducting plates separated by an insulator; thus, a steady current *cannot* pass through the capacitor. (See the previous sections on capacitors.) A time-varying current can add or remove charges from the capacitor plates. A simple circuit for charging a capacitor is shown in Figure 4-28.

Figure 4-28 An RC circuit for charging a capacitor.

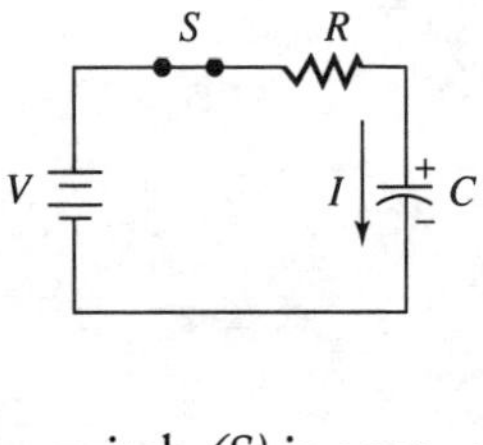

Initially, at time $t = 0$, the switch *(S)* is open, and there is no charge on the capacitor. When the switch is closed, a current will pass through the resistor and charge the capacitor. The current will cease when voltage drop across the capacitor equals the potential of the battery *(V)*. Once the capacitor reaches the maximum charge, the current will decrease to zero. The current is at maximum immediately after the switch is closed and decreases exponentially with time. The **capacitive time constant** (τ), the Greek letter tau) is the time for the charge to decay to $1/e$ of its initial value, where e is the natural logarithm. A capacitor with a large time constant will change slowly. The capacitive time constant is $\tau = RC$.

From Kirchhoff's rules, the following expressions for the potential difference across the capacitor *(V_C)* and the current *(I)* in the circuit are derived:

$$V_C = V\left(1 - e^{\frac{-t}{RC}}\right) \text{ and } I = \frac{V}{R} e^{\frac{-t}{RC}}$$

where V is the potential of the battery.

Resistor-inductance circuits

A circuit with a resistor, an inductor, and an ac generator is an **RL circuit.** When the switch is closed in an RL circuit, a back emf is induced in the inductor coil. (See the section "Mutual inductance and self-inductance.") The current, therefore, takes time to reach its maximum value, and the time constant, called the **inductive time constant,** is given by

$$\tau = \frac{L}{R}$$

The equations for the current as a function of time and for the potential across the inductor are

$$I = \frac{V}{R}\left(1 - e^{\frac{-Rt}{L}}\right) \text{ and } V_L = Ve^{\frac{-Rt}{L}}$$

A switch was used in the above discussions of RC and RL circuits for simplicity. Opening and closing a switch gives a response similar to that of an ac current. The RC and RL circuits are similar to each other because an increase in voltage yields a current that changes exponentially in each circuit, but the responses are different in other ways. These different behaviours, described below, lead to different responses in ac circuits.

Reactance

Now consider an ac circuit consisting only of a capacitor and an ac generator. The plots of current and voltage across the capacitor as a function of time are shown in Figure 4-29. The curves are *not* in phase as they were for the circuit of a resistor and an ac generator. (Refer to Figure 4-27.) The curves indicate that, for a capacitor, the voltage reaches its maximum value one quarter of a cycle after the current reaches its maximum value. Thus, the voltage *lags* the current through the capacitor by 90 degrees.

Figure 4-29 Current and voltage from an ac source through a capacitor.

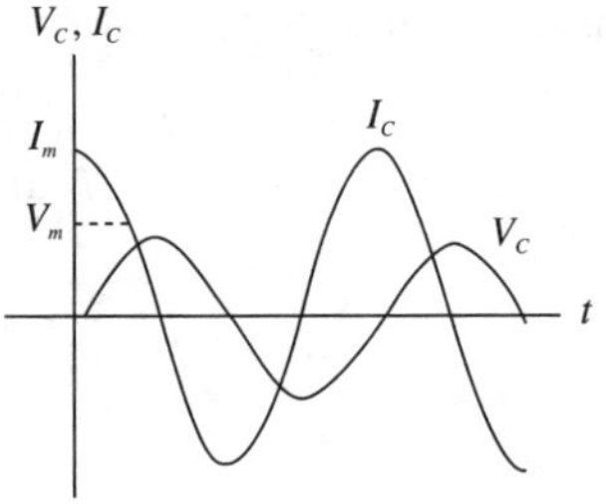

The **capacitive reactance** (X_C) expresses the impeding effect of the capacitor on the current and is defined as

$$X_C = \frac{1}{2\pi f C}$$

where C is in farads and the frequency (f) is in units of hertz. Ohm's law yields $V_C = IX_C$, where V_C is the rms voltage across the capacitor and I is the rms current in the circuit.

Consider a circuit with only an inductor and an ac generator. Figure 4-30 shows the plots of the current and voltage as a function of time for the inductor. Note again that the voltage and current are not in phase. The voltage for this circuit reaches its maximum value one quarter of a cycle before the current reaches its maximum; thus, the voltage *leads* the current by 90 degrees.

Figure 4-30 Current and voltage from an ac source through an inductor.

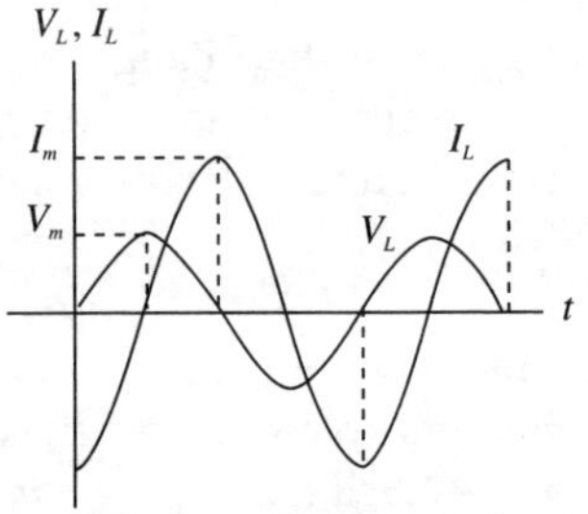

The current in the circuit is impeded by the back emf of the inductor coil. The effective resistance is called the **inductive reactance** (X_L) defined by $X_L = 2\pi fL$, where L is measured in henries and f is in hertz. Ohm's law yields $V_L = IX_L$, where V_L is the rms voltage across the inductor and I is the rms in the inductor.

Resistor-inductor-capacitor circuit

A circuit with a resistor, an inductor, a capacitor, and an ac generator is called an **RLC circuit**. The phase relationships of these elements can be summarized as follows:

- The instantaneous voltage across the resistor (V_R) is in phase with the instantaneous current.
- The instantaneous voltage across the inductor (V_L) leads the instantaneous current by 90 degrees.
- The instantaneous voltage across the capacitor (V_C) lags the instantaneous current.

Because the voltages across the different elements are not in phase, the individual voltages cannot be simply added in ac circuits. The equations for the total voltage and the phase angle are

$$V = \sqrt{V_R^2 + (V_L - V_C)^2} \text{ and } \tan\theta = \frac{V_L - V_C}{V_R}$$

where all voltages are rms values. Ohm's law for the general case of ac circuits is now expressed $V = IZ$, where R is replaced by **impedance** (Z), measured in ohms. The impedance is defined as

$$Z = \sqrt{R^2 + (X_L - X_C)^2}$$

Power

The resistor is the only element that dissipates power in an ac circuit. The average power is given by $P_{avg} = I^2R$. The **power factor,** the average power dissipated in the ac circuit, is given by $P_{avg} = IV \cos\theta$.

Resonance

From the general expression for Ohm's law of the ac circuit, the current can be written

$$I = \frac{V}{Z} = \frac{V}{\sqrt{R^2 + (X_L - X_C)^2}}$$

The equation indicates that the current has its maximum value when the impedance has its minimum value. The frequency for this situation is called the **resonant frequency.** It is defined as

$$f_o = \frac{1}{2\pi\sqrt{LC}}$$

Transformers

A transformer is an ac version of the induction coil applied to transform voltages from one circuit to another. The device is constructed of two coils with differing number of turns wrapped on a common iron core. In Figure 4-31, assume that V_1 is the input (primary) voltage and V_2 is the output (secondary) voltage.

Figure 4-31 A simple transformer.

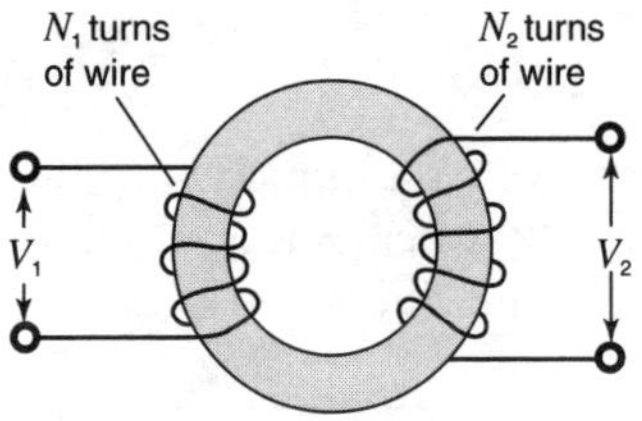

The ratio of input to output voltage is

$$\frac{V_1}{V_2} = \frac{N_1}{N_2}$$

where N_1 and N_2 are the respective number of turns of wire of each coil. Depending upon the number of turns in each coil, the transformer can increase (step up) or decrease (step down) the voltage. This equation is for an ideal transformer. Actual transformers have efficiencies of 90 to 99 per cent.

Chapter Checkout

1. Three point charges are arranged on an *x-y* coordinate system in a triangle: $q_1 = 0.5$ C located 15 m from the origin along the $+x$ axis; $q_2 = -1.0$ C located 20 m from the origin along the $+y$ axis; and $q_3 = +0.25$ C located 18 m from the origin along the $-x$ axis. Calculate the force felt by q_2. Be sure to calculate magnitude and direction.
2. Using Gauss's law, determine the electric field at a distance r from an infinite rod of charge which has a positive linear charge density λ ($\lambda = \Delta Q/\Delta L$, that is, charge per unit length of the rod, measured in C/m).
3. The membrane of a cardiac cell may be modeled as a simple RC circuit, with the resistor, capacitor, and battery in parallel with each other. Assume the cell maintains a potential difference across the membrane such that the inside is –60 mV (the exterior is taken to be 0 V). If the dielectric constant of the membrane is 5 and the capacitance per unit area is 1 μ/cm^2: **(a)** How thick is the membrane? **(b)** What is the energy stored in a typical cell, assuming it is a sphere with radius 10 μm? **(c)** How much total charge is separated by the membrane?
4. An electron (charge $-e$) traveling through space with velocity $v = 1$ km/s moves parallel to an infinite wire carrying a current of 3 mA. If the electron moves in the **(a)** same direction or **(b)** opposite direction as the current in the wire, calculate the force it feels due to the wire's magnetic field, when it is 1 m from the wire. Be sure to note the direction of the force in each case. **(c)** How do your answers change for a positron (charge $+e$)?
5. Using Gauss's law, calculate the electric field **(a)** inside and **(b)** outside a solid sphere with radius R and uniform positive internal charge density ρ (charge density is $\rho = Q/V$, that is, charge per unit of volume, measured in C/m^3).

Answers: 1. 8.4×10^6 N, 15.6° from the $+y$ axis toward q_1 **2.** $E = \lambda/2\pi\varepsilon_o r$ pointing away from the rod **3a.** 4.4 nm **b.** 2.26×10^{-14} J **c.** 7.5×10^{-13} C **4a.** 9.6×10^{-26} N, away from the wire **b.** 9.6×10^{-26} N, toward the wire **c.** The directions would reverse **5a.** $E = \rho r/3\varepsilon_o$, out from the centre **b.** $E = \rho R^3/3\varepsilon_o r^2 = Q/4\pi\varepsilon_o r^2$, out from the centre

Chapter 5
LIGHT

CHAPTER CHECK-IN

- ❑ Understanding the basic characteristics of light
- ❑ Applying the rules of geometric optics to understand mirrors, lens, and refraction
- ❑ Analyzing the wave properties of light to understand diffraction and interference

Newton proposed the **particle theory of light** to explain the bending of light upon reflection from a mirror or upon refraction when passing from air into water. In his view, light was a stream of particles emitted from a light source, entering the eye to stimulate sight. Newton's contemporary Christiaan Huygens showed that a **wave theory of light** could explain the laws of reflection and refraction. In the late 1800s, James Clerk Maxwell predicted, and then Gustav Ludwig Hertz verified, the existence of electromagnetic waves traveling at the speed of light. A complete conceptualization of the nature of light includes light as a particle, as a wave, and as electromagnetic radiation.

Characteristics of Light

The modern view is that light has a dual nature. To debate whether light is a particle or a wave is inappropriate because in some experiments light acts like a wave and in others it acts like a particle. Perhaps it is most accurate to say that both waves and particles are simplified models of reality and that light is such a complicated phenomena that no one model from our common experience can be devised to explain its nature.

Electromagnetic spectrum

Maxwell's equations united the study of electromagnetism and optics. Light is the relatively narrow frequency band of electromagnetic waves to which our eyes are sensitive. Figure 5-1 illustrates the spectrum of **visible light.** Wavelengths are usually measured in units of nanometres ($1 nm = 10^{-9} m$) or in units of angstroms ($1Å = 10^{-10} m$). The colours of the visible spectrum stretch from violet, with the shortest length, to red, with the longest wavelength.

Figure 5-1 The spectrum of electromagnetic radiation, which includes visible light.

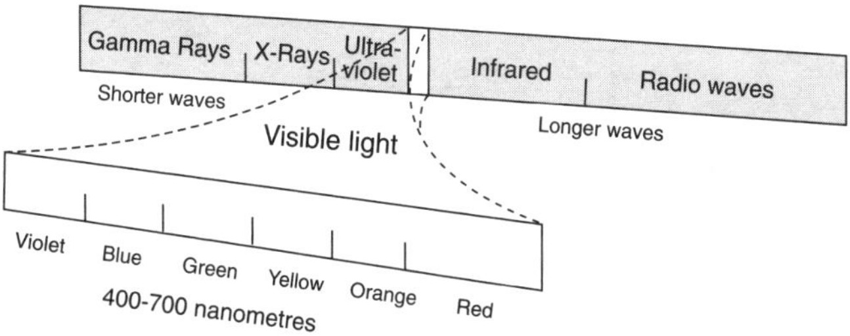

Speed of light

Light travels at such a high speed, 3×10^8 m/sec, that historically it was difficult to measure. In the late 1600s, Claus Roemer observed differences in the period of the moons of Jupiter, which varied according to the position of the earth. He correctly assumed a finite speed of light. He deduced the annual variation was due to a changed distance between Jupiter and the earth; so a longer period indicated that the light had farther to travel. His estimate, 2.1×10^8 m/s, based on his value for the radius of the earth's orbit, was inaccurate, but his theories were sound. Armand Fizeau was the first to measure the speed of light on the earth's surface. In 1849, he used a rotating toothed wheel to find a close approximation of the speed of light, 3.15×10^8 m/s. As shown in Figure 5-2, a light beam passed through the wheel, was reflected by a mirror a distance *(d)* away, and then again passed through an opening between cogs.

Figure 5-2 Fizeau's apparatus for measuring the speed of light.

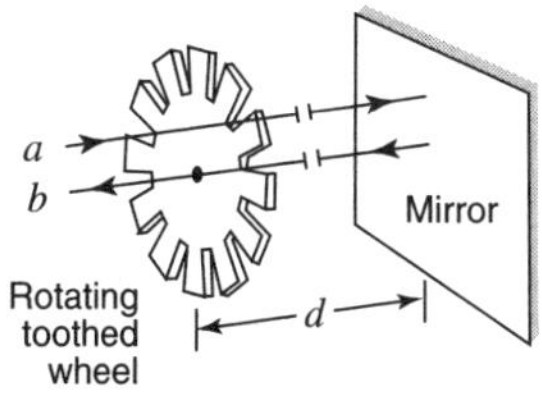

Assume the speed of the wheel is adjusted so that the light passing through the opening *a* then passes through opening *b* after reflection. If the toothed wheel spins at an angular velocity ω and the angle between the two openings is θ, then the time for light to travel 2*d* is

$$t = \frac{\theta}{\omega}$$

and so the velocity of light may be calculated from

$$c = \frac{2d}{t} = \frac{2\omega d}{\theta}$$

where *c* denotes the speed of light. More modern methods with lasers have made measurements accurate to at least nine decimal places.

Polarization

Light and other elecromagnetic radiation can be polarized because the waves are transverse. Recall from the wave motion section in Chapter 2 that an oscillatory motion perpendicular to the direction of motion of the wave is the distinguishing characteristic of transverse waves. Longitudinal waves, such as sound, cannot be polarized. **Polarized** light has vibrations confined to a single plane that is perpendicular to the direction of motion. A beam of light can be represented by a system of light vectors. In Figure 5-3, unpolarized light is radiating from a light bulb. The beam going to the top of the page is viewed along the direction of motion (as end-on). The vectors in the beam traveling to the side of the page are seen perpendicular to the direction of motion (as a side view).

Figure 5-3 A light bulb emits unpolarized light.

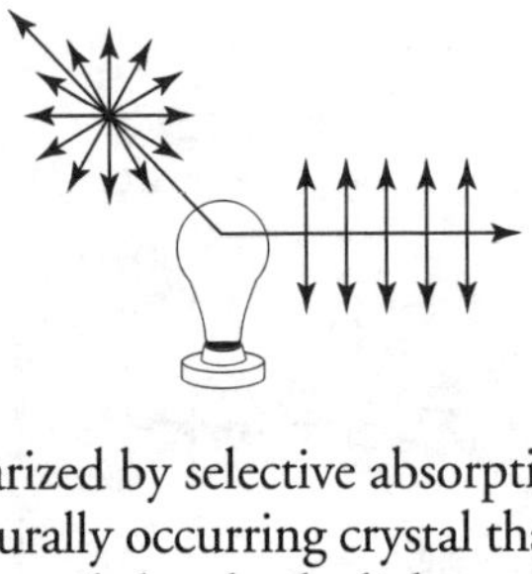

Light is commonly polarized by selective absorption of a polarizing material. Tourmaline is a naturally occurring crystal that transmits light in only one plane of polarization and absorbs the light vectors in other polarization planes. This type of material is called a **dichroic** substance. A mechanical analogy illustrates this process. Imagine a rope with transverse pulses passing through two frames of slots, as shown in Figure 5-4. When the second polarizer is turned perpendicular to the first, the wave energy is absorbed.

Figure 5-4 A mechanical analogy of polarization, for a wave on a string.

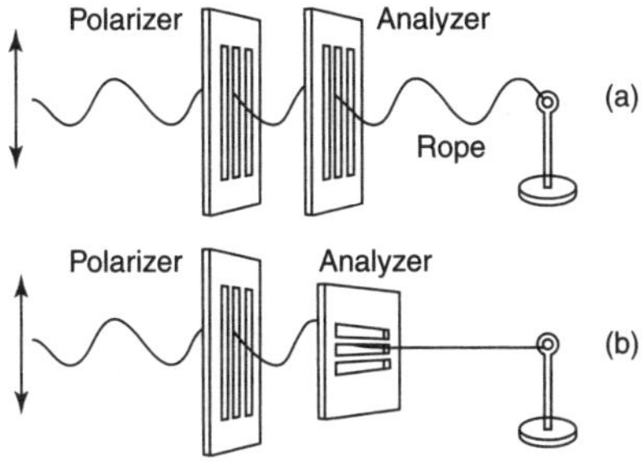

Polaroid, another dichroic substance, is manufactured from long-chain hydrocarbons with alignment of the chains. As you will recall, electromagnetic waves are crossed electric and magnetic fields propagating through space. The orientation of the electric wave is taken as the direction of polarization. The polaroid molecules can conduct electric charges parallel to their chains; therefore, hydrocarbon molecules in polaroid filters absorb light with an electric field parallel to their length and transmit light with electric field perpendicular to their length.

Figure 5-5 shows the direction of light vectors for a beam of light traveling through two polaroids. The first polaroid is called the **polarizer,** and the second polaroid is called an **analyzer.** When the transmission axes of the polarizing materials are parallel, the polarized light passes through. Light is nearly completely absorbed when passing through two sets of polarizing materials with their transmission axes at right angles.

Figure 5-5 A sequence of polaroids.

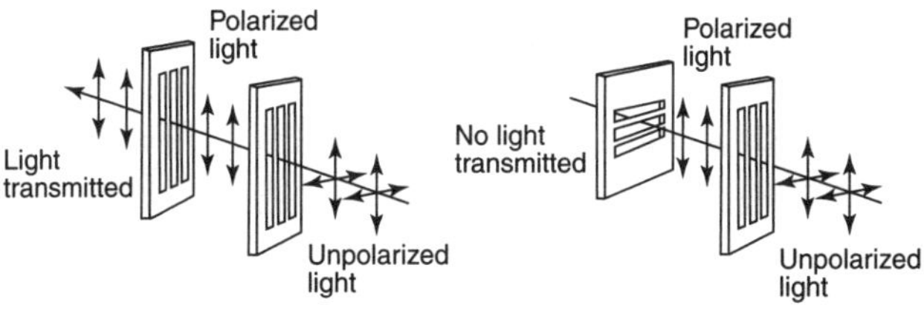

Light can be polarized by reflection. For this reason, polaroid sunglasses are effective for reducing glare. Sunlight is primarily polarized parallel to the surface after reflection; therefore, the polaroids in sunglasses are oriented so that the reflected polarized light is largely absorbed.

Geometrical Optics

When an object is dropped in still water, the circular wave fronts that are produced move out from the contact point over the two-dimensional surface. A light source emits light uniformly in all directions of the three-dimensional world. The wave fronts are spherical, and the direction of motion of the wave is perpendicular to the wave front, as depicted in Figure 5-6. This straight line path shown by the arrow is called a **ray.** Depicting light as rays in **ray diagrams** provides a method to explain the images formed by mirrors and lenses.

Figure 5-6 Rays are perpendicular to the spherical wave fronts.

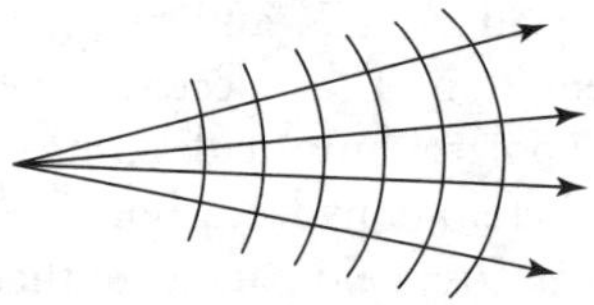

Far from the source, the curvature of the wave front is small, so the wave front appears to be a plane. Then, the light rays will be nearly parallel. Rays from the sun are considered to be parallel when reaching the earth.

The law of reflection

Most visible objects are seen by reflected light. There are few natural sources of light, such as the sun, stars, and a flame; other sources are man-made, such as electric lights. For an object to be visible, light from a source is reflected off the object into our eyes (except in the special case of phosphors). In Figure 5-7, the light is coming from the sun, parallel due to the distance of the source. The light reflects off the object and travels in straight lines to the viewer. Through experience, the viewer has learned to extend the reflected rays entering the eye back to locate the object.

Figure 5-7 Vision is the result of light reflected from the object.

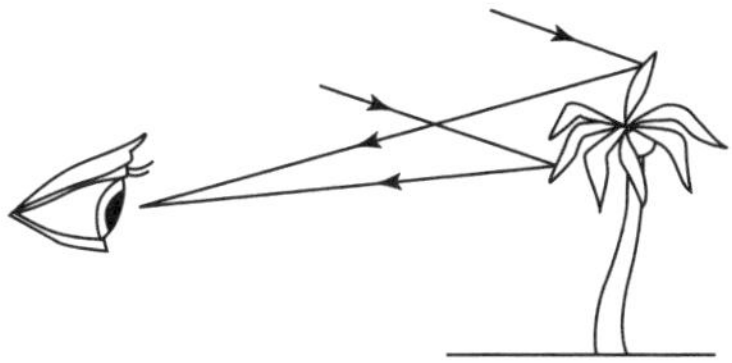

As shown in Figure 5-8, light strikes a mirror and is reflected. The original ray is called the **incident ray,** and after reflection, it is called the **reflected ray.** The angles of the incident and reflected rays are always measured from the **normal.** The normal is a line perpendicular to the surface at the point where the incident ray reflects. The incident ray, reflected ray, and normal all lie in the same plane perpendicular to the reflecting surface, known as the **plane of incidence.** The angle measured from the incoming ray to the normal is termed the **incident angle.** The angle measured from the outgoing ray to the normal is called the **reflected angle. The law of reflection** states that the angle of incidence equals the angle of reflection. This law applies to all reflecting surfaces.

Figure 5-8 The law of reflection.

Light undergoes either diffuse or regular reflection. The two are illustrated in Figure 5-9.

Diffuse reflection occurs when light reflects from a rough surface. **Regular reflection** is reflection from a smooth surface, such as a mirror. The reflected rays are scattered in diffuse reflection. This scattering is because the local direction of the normal to the surface is different for the different rays. By contrast, in regular reflection, the reflected light rays are orderly because each local region of the surface has a normal in the same direction.

Figure 5-9 (a) Regular reflection. (b) Diffuse reflection.

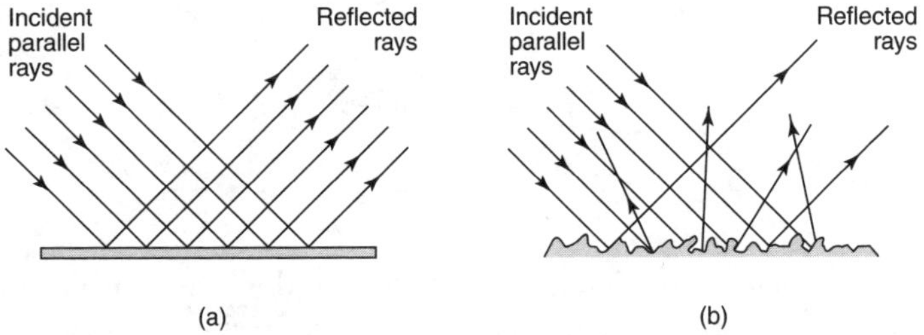

Plane mirrors

Figure 5-10 illustrates the formation of an image by a plane mirror. Light rays are coming from a source and reflecting off each point of the object *(AB)* in all directions. For simplicity, only a few of the rays are drawn. The rays spread upon leaving the object, and then each ray reflects from the mirror according to the law of reflection. The eye extends back the diverging reflected rays to see an image behind the mirror. An image formed in this manner by extending back the reflected diverging rays is called a **virtual** image. A virtual image cannot be projected on a screen. The light does *not* physically come together, but rather, the eye (or camera) interprets the diverging rays as originating from an image behind the mirror. Due to the law of reflection, the image formed by a plane mirror is the same distance behind the mirror as the object is in front of the mirror.

Figure 5-10 Construction of an image reflected in a mirror.

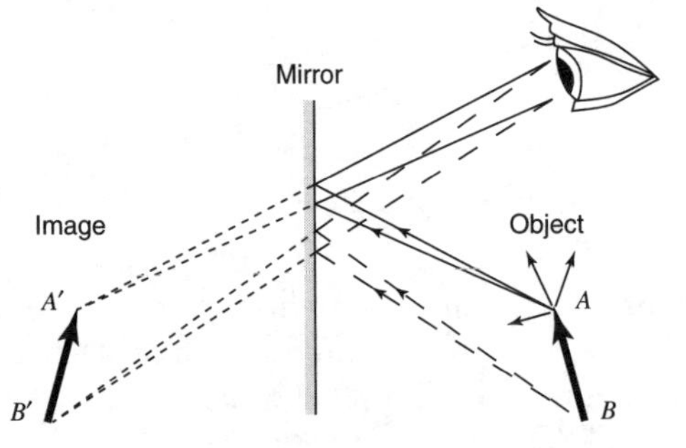

How tall does a mirror need to be so you can see your entire height? Assume the top of the mirror is in line with the top of your head. Does it matter where you stand? The ray diagram in Figure 5-11 illustrates this situation.

From the law of reflection and basic geometry, it can be proven that the marked angles are all equal; therefore, the necessary height of the mirror is approximately half your height. Draw a figure at a different distance to show that the distance from the person to the mirror does not change the result.

Figure 5-11 Seeing your own feet in a mirror.

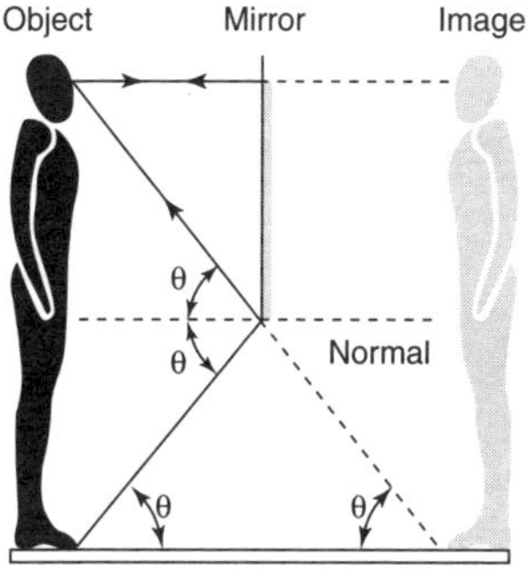

Concave mirrors

Regular reflection occurs not only for plane (flat) mirrors but also for curved mirrors. Picture a series of plane mirrors arranged in a semicircle as shown in Figure 5-12. The incoming light is from a distant source and, therefore, is nearly parallel, as Figure 5-12(a) shows. After reflection, the light converges on a region. As the number of mirrors increases—Figure 5-12(b)—the converging region of the light beams decreases.

A **concave** mirror reflects its light from the inner curved surface. The mirror can be a portion of a sphere, a cylinder, or shaped as a rotated parabolic curve. The light rays intersect after reflection at a common focus called the **focal point** *(F)*. The focal point is on the **optical axis,** the symmetry axis of the mirror. The distance f from focal point to the mirror is called the **focal length.** For a spherical mirror, the focal length is one-half the radius of the sphere that defines the mirror. This distance c is called the **radius of curvature,** and the centre of the sphere is denoted as C ($c = 2f$). Figure 5-13 illustrates these definitions.

Figure 5-12 A semicircular arrangement of mirrors focuses light in the region *F*.

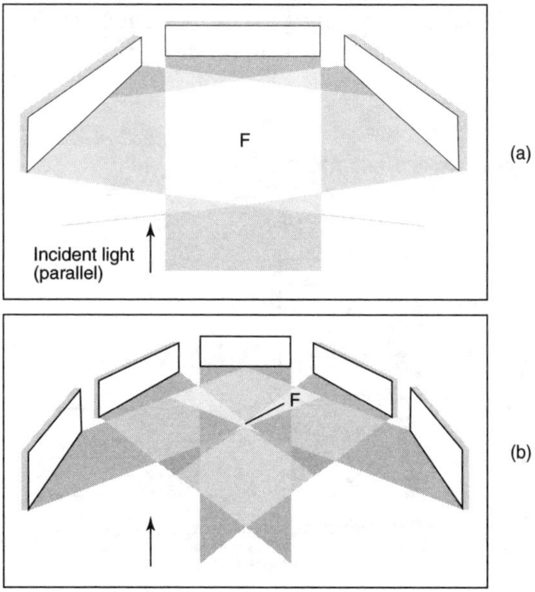

Figure 5-13 A concave mirror with radius *c* and focal length *f*.

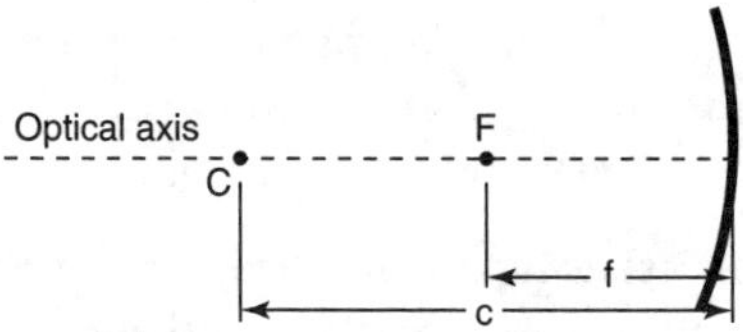

It is helpful to have a geometric system for locating an image formed by rays reflected from a curved mirror. Any reflected ray follows the law of reflection; however, certain rays have easily defined paths so that measuring angles and finding the normals are not necessary. Four of these rays are

- The ray directed parallel to the optical axis will reflect through *F*.
- The ray directed through *F* will reflect parallel to the optical axis.
- The ray directed to the centre of the mirror will reflect at the same angle to the optical axis.
- The ray directed along the radius of the sphere will reflect back on itself.

Light rays are drawn for four different positions in Figure 5-14: (a) far from *F*, (b) at nearly 2*F*, (c) between *F* and 2*F*, and (d) at *F*. It is only necessary to find the intersection of two reflected rays from a point on the object to define the corresponding point on the image. A third one can be used as a check. Sometimes one or another of the rays may be difficult to draw, and so choices can be made.

Figure 5-14 Images from a concave mirror.

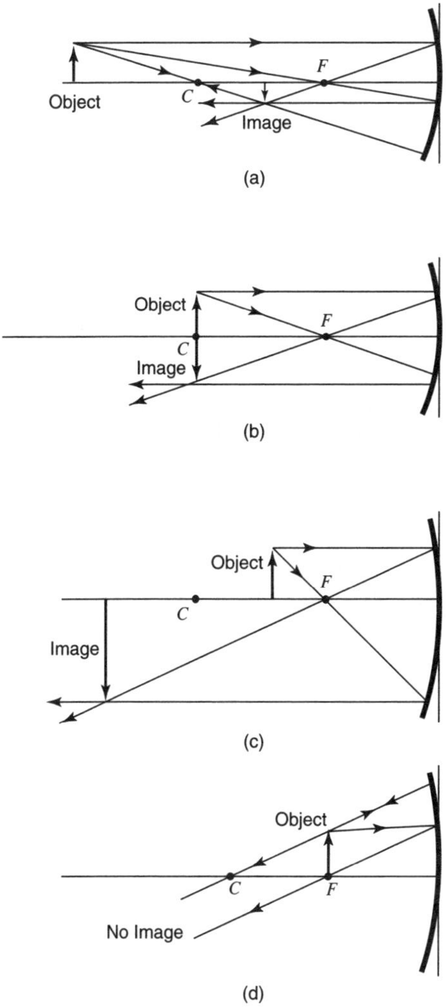

Notice that images are formed for the first three cases but not for the last one. No image is formed when the object is at the focal point or, alternatively, the image is formed at infinity and cannot be seen. The three images are all **real images.** Real images can be shown on a screen because the light physically comes together at a point in space. Note that real images are formed by light that converges after reflection. Also, real images are always inverted—upside down—with regard to the original object. In Figure 5-14, the light rays from the bottom of the object are not drawn. Light traveling along the optical axis will reflect back along the axis, and so if a point of the object is on the optical axis, the corresponding image point will also be on the optical axis.

The images formed can be characterized by size and placement. Let the distance from the object to the mirror be given by O. Then the image characteristics can be summarized as follows:

- If $O > 2F$, the image is inverted, smaller, and located between F and $2F$.
- If $O = 2F$ (at C), the image is inverted, the same size as the object, and located at $2F$; that is, the distances of both the object and image to the mirror are equal.
- If $2F < O < F$, the image is inverted, larger than the object, and located $>2F$. Light paths are reversible. If the object is placed in the position of its former image, the image will then be located where the object was originally; that is, the two will exchange positions.

Figure 5-15 shows the diagram for the case when the object is between the focal point *(F)* and the mirror. In this case, a **virtual image** is formed because the reflected rays diverge from the surface of the mirror. The virtual image is upright, enlarged, and behind the mirror. Virtual images are *never* inverted.

Figure 5-15 Formation of a virtual image in a concave mirror.

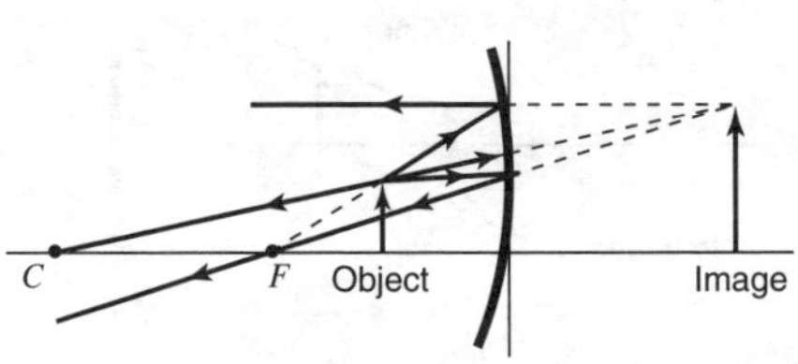

The following approximate **mirror equation** relates the distances from the object to the mirror *(O)*, the distance from the image to the mirror *(I)*, and the focal length *(f)*:

$$\frac{1}{O}+\frac{1}{I}=\frac{1}{f}$$

The sign of f is positive if it is on the same side as the mirror (a concave mirror) and negative otherwise (convex mirror). Both O and I are positive in sign if they lie on the same side of the mirror as the incident light and negative if they lie on the opposite side.

The magnification is defined as the ratio of the image size to the object size. This ratio is the same as the ratio of the distances:

$$\text{magnification}=\frac{I}{O}$$

Thus, a magnification of 10× means the image seen is 10 times the size of the object when viewed without a magnifying device.

Convex mirrors

The graphical technique for locating the image of a convex mirror is shown in Figure 5-16. For convex mirrors, the image on the opposite side of the mirror is virtual, and the images on the same side of the mirror are real. Figure 5-16 shows a virtual, upright, and smaller image. In comparison to the virtual image of the concave mirror, the virtual image of the convex mirror is still upright, but it is diminished (smaller) instead of enlarged and on the opposite side of the mirror instead of the same side. Again, the virtual image is formed by extending back the reflected diverging rays.

Figure 5-16 Formation of a virtual image in a convex mirror.

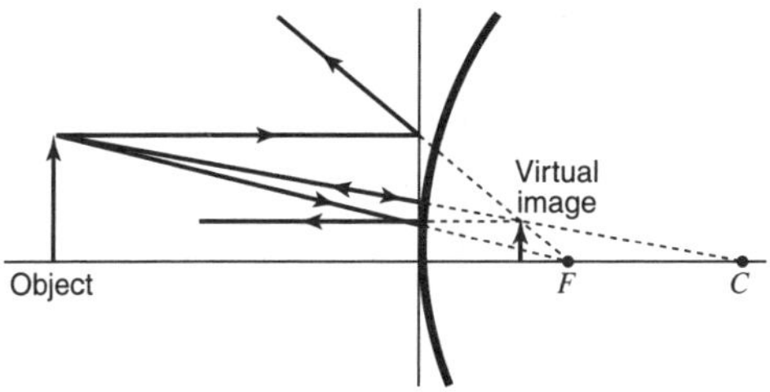

The law of refraction

Refraction is the bending of light when the beam passes from one **transparent** medium into another. A transparent object allows the transmission of light, in contrast to an **opaque** object, which does not. Some of the light will also be reflected. The incident ray, reflected ray, normal, and **refracted ray** are shown in Figure 5-17.

Figure 5-17 The law of refraction.

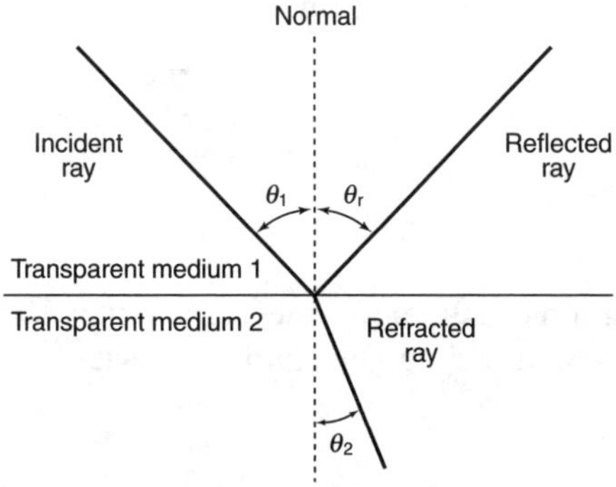

When Willebrod Snell (1580–1626) observed light traveling from air into another transparent material, he found a constant ratio of the sines of the angles measured from the normal to the light ray in the material:

$$n = \frac{\sin \theta_{air}}{\sin \theta_{material}}$$

The constant *(n)* is called the **index of refraction** and depends only upon the optical properties of the material. The index of refraction gives a measure of the amount of bending occurring when light travels from air into the material. It is a dimensionless number and can be located in tables of properties of materials. For example, the index of refraction of water is 1.33, and the index of refraction of crown glass varies from 1.50 to 1.62, depending upon the composition of the glass.

For the more general case of light traveling from medium 1 to medium 2, **Snell's law** can be written $n_1 \sin \theta_1 = n_2 \sin \theta_2$, where the subscripts 1 and 2 refer respectively to the angles and indices of the refraction for material 1 and material 2 respectively. A light ray traveling along the normal, with an incident angle of zero, will not be bent.

The index of refraction is also the ratio of the speed of light in a vacuum (*c*) and the speed of light in that medium (*v*); thus,

$$n = \frac{c}{v}$$

Consider the following problem involving both reflection and refraction. Imagine light entering an aquarium and reflecting off a mirror at the bottom. First, what will be the angle of refraction in the water if the angle in air is 30 degrees? Second, at what angle will the beam leave the water? See the setup in Figure 5-18.

Figure 5-18 A problem combining refraction and reflection.

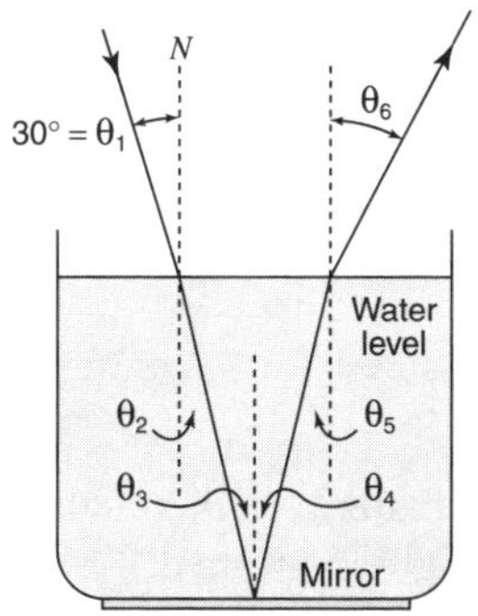

Angle θ_2 is determined from θ_1, using Snell's law of refraction. Angle $\theta_2 = \theta_3$ by geometry, $\theta_3 = \theta_4$ by law of reflection, and $\theta_4 = \theta_5$ by geometry. θ_6 is related to θ_5 by Snell's law of refraction, in the same ration as θ_1 toθ_2. Therefore, θ_6—the angle of the ray leaving the water—must be 30 degrees. The problem is symmetrical.

A light ray passing through a rectangular block of transparent material will simply be displaced from its original path. For example, in passing from air to glass, the ray will bend toward the normal. Upon leaving the glass block, the ray will bend away from the normal so that the measured angles in the air on each side of the block are the same (Figure 5-19).

Figure 5-19 A light ray is displaced after passing through a refracting medium.

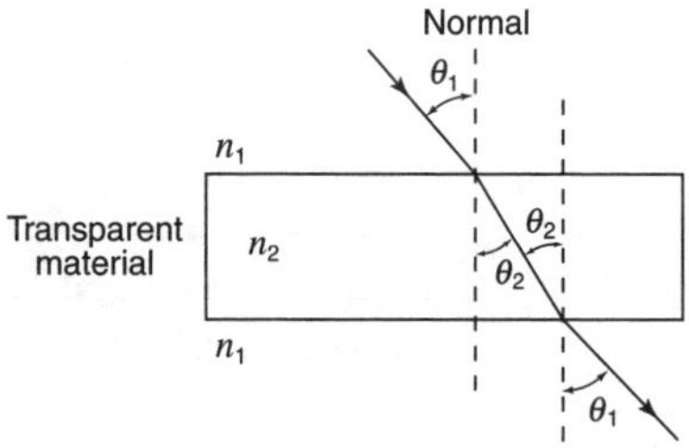

Brewster's angle

In the earlier section "Polarization," it was stated that light reflected from the surface of a material is partially polarized. A ray incident on a transparent surface at a certain angle will be partly refracted and partly reflected in a plane polarized ray. This angle of maximum plane polarization is called **Brewster's angle,** named for Sir David Brewster (1781–1868). The equation is $\tan \theta = n$, where n is the index of refraction of the reflecting surface.

Total internal reflection

When light travels from a material with a higher n to one with a lower n, at certain angles all of the light is reflected. This effect is called **total internal reflection.**

Example 1: Figure 5-20 illustrates ray 1 along the normal (no bending), rays 2 and 3 are refracted, and rays 5 and 6 are reflected. Ray 4 is intermediate between reflection and refraction with an angle of refraction of 90 degrees. The incident angle for this case is called the **critical angle** (θ_c). If the angle of incidence is less than θ_c, the light will refract, and if it is greater, the light will reflect.

Figure 5-20 Total internal reflection at the interface of two different media.

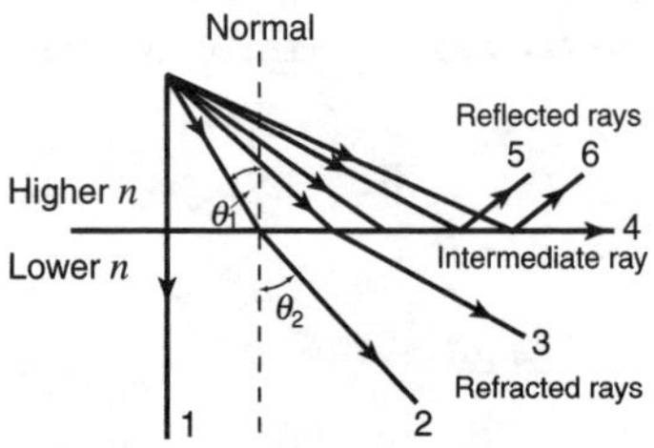

The equation is

$$\sin\theta_c = \frac{n_2}{n_1}$$

where $n_1 > n_2$. Find the critical angle from glass to air.

Solution: $\theta_c = \sin^{-1}\left(\frac{n_2}{n_1}\right) = \sin^{-1}\left(\frac{1.00}{1.52}\right)$

$\theta_c = 41.1^\circ$

Therefore, if the incident ray on a glass to air interface is greater than 42 degrees, total internal reflection will occur. Figure 5-21 shows the light rays entering and leaving a 45-45-90 glass prism. This phenomenon has broad applications where a mirror is needed, but a silvered surface might corrode after a period of time.

Figure 5-21 Total internal reflection in a glass prism.

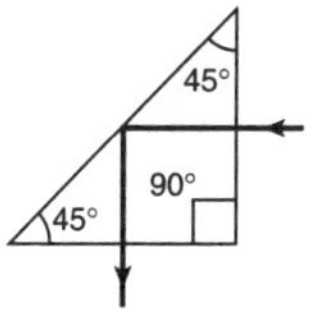

Optical lenses

An optical lens functions by refracting light at its interfaces. In these examples, the lens will be assumed to be thin, in which case the thickness of the lens is negligible compared with it focal length. Lenses are basically of two types. A **converging lens** causes parallel rays to converge, and a **diverging lens** causes parallel rays to diverge. Figure 5-22 illustrates the paths of the rays through the lens and the focal point for each case. The definitions for optical axis, focal point, and focal length given for curved mirrors hold true for lenses, with the addition that lenses have focal points on each side of the lens.

Ray diagrams can be made for lenses similar to those drawn for curved mirrors. These three rays can be drawn to locate the image formed by the lens.

- The ray directed parallel to the optical axis refracts through F on the far side.
- The ray directed to the near F refracts parallel to the optical axis.
- The ray directed to the centre of the lens is undeviated (in the thin lens approximation).

Figure 5-22 Thin lenses function by refracting light.

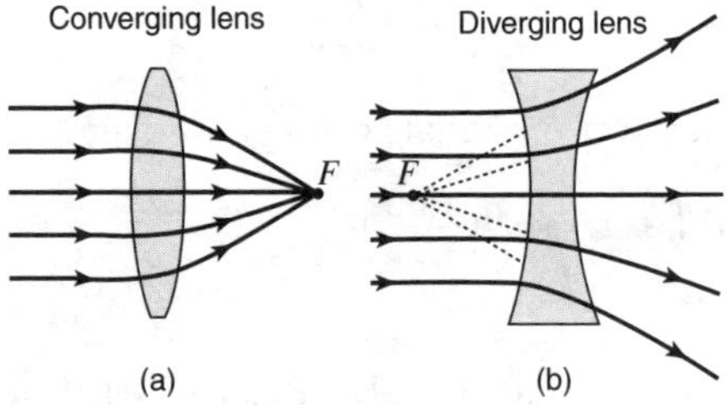

The ray diagrams for two cases of a converging lens are shown in Figure 5-23.

Figure 5-23 Ray diagrams for a converging lens, showing the formation of (a) a real image or (b) a virtual image.

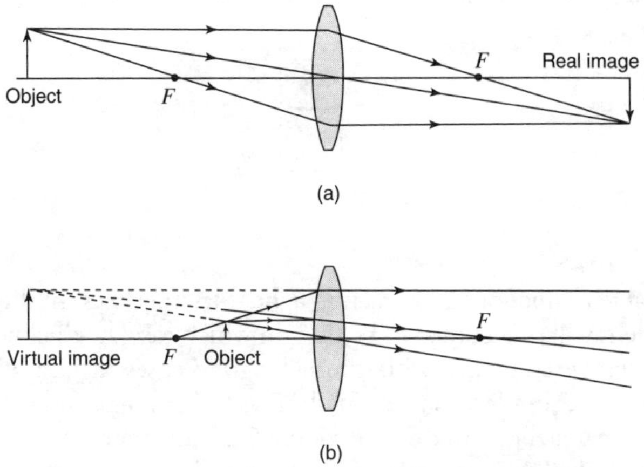

In Figure 5-23(a), a real image is formed, and in Figure 5-23(b), a virtual image is formed. The lens setup in Figure 5-23(b) is called a simple magnifier. With lenses as with mirrors, virtual images are right side up, and real images are inverted. (This is why slides inserted into a projector are inverted; the projector lens reinverts the image on the screen.)

The **lens equation** is the same relationship used for curved mirrors:

$$\frac{1}{O} + \frac{1}{I} = \frac{1}{f}$$

as is the equation for magnification:

$$\text{magnification} = \frac{I}{O}$$

The focal length is positive for a converging lens and negative for a diverging lens. The object and image distances are positive if they are on opposite sides of the lens and negative if they are on the same side. The relative sizes and positions of the object and image for a converging lens are similar to the four cases reviewed for the concave mirror.

- If $O > 2F$, the image is inverted, smaller, and located between F and $2F$, on the opposite side.
- If $O = 2F$, the image is inverted, the same size as the object, and located at $2F$; that is, the distances of both the object and image to the lens are equal but on opposite sides of the lens.
- If $2F < O < F$, the image is inverted, larger than the object, and located $> 2F$.
- If $O < F$, the image is virtual, enlarged, and located on the same side of the lens where $I > F$.

Figure 5-24 shows the ray diagram for a diverging lens. The image formed by this lens is always virtual, upright, and diminished.

Figure 5-24 Ray diagram for a diverging lens.

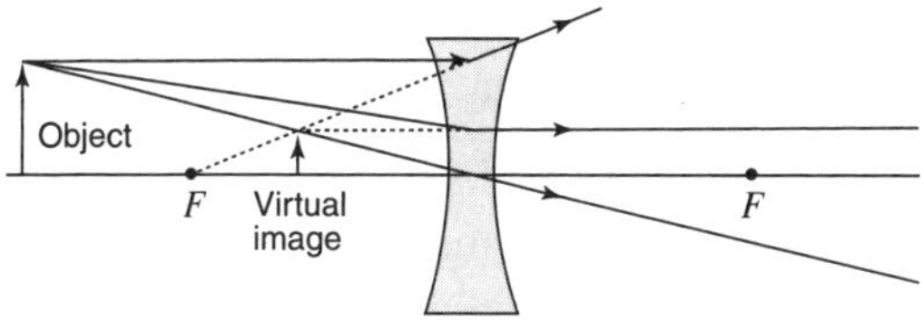

The compound microscope

When lenses are used in combinations, the image given by one lens becomes the object for a second lens. The **compound microscope** is an example of the use of several lenses to magnify an object. An **objective** lens near the object forms an enlarged image. This image is then further magnified by the second lens, called the **eyepiece.** Both are converging lenses.

In Figure 5-25, the object *(AB)* is placed just below the focal point of the objective lens. The objective lens forms an enlarged, real, and inverted image at a distance greater than $2f$ from the first lens. This image $(A'B')$ falls inside the focal point of the eyepiece lens; therefore, an enlarged, virtual image is formed by the eyepiece $(A''B'')$. The total magnification is the product of the magnifications of each lens.

Figure 5-25 A compound microscope.

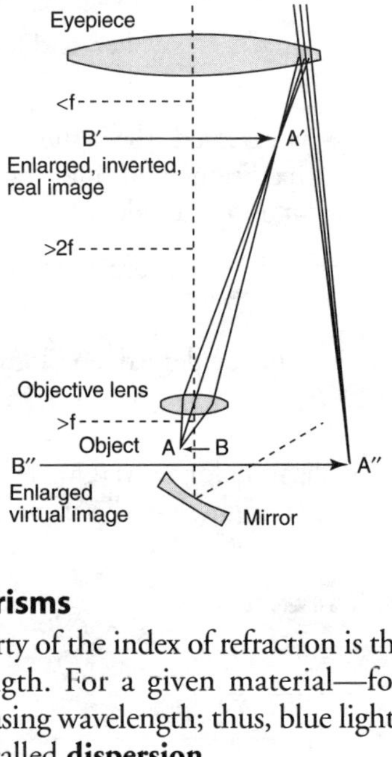

Dispersion and prisms

An important property of the index of refraction is that it is slightly dependent upon wavelength. For a given material—for example, glass—n decreases with increasing wavelength; thus, blue light bends more than red light. This effect is called **dispersion.**

Light is refracted twice as it enters and leaves the prism, as shown in Figure 5-26.

Figure 5-26 Dispersion of white light in a prism.

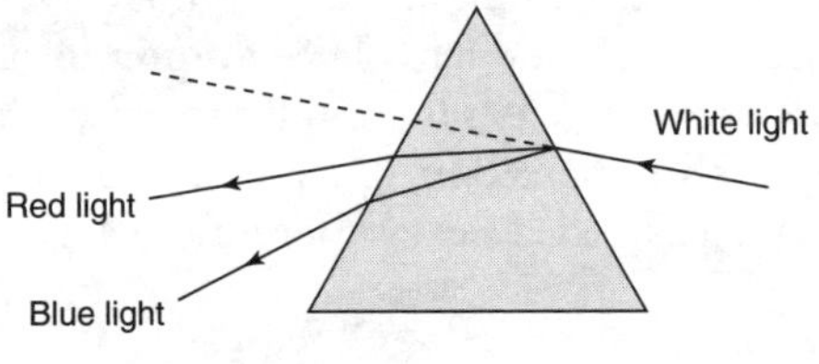

A given ray is bent from its original direction of travel by an angle (δ), called the **angle of deviation.** The angle of deviation for the red light is less than that for the blue light; therefore, the prism spreads the light into the colours of the **spectrum.** These colours are commonly called red, orange, yellow, green, blue, indigo, and violet (often abbreviated with the mnemonic Roy G Biv).

Rainbows are formed by dispersion and total internal reflection of sunlight in raindrops. The critical angle for water to air is approximately 40 degrees. The sunlight enters the drop and is reflected off the side of the drop away from the viewer. Due to dispersion, the violet ray emerges above the red ray. Figure 5-27 shows the refraction of sunlight on one idealized raindrop.

Figure 5-27 Dispersion and internal reflection of light passing through a raindrop.

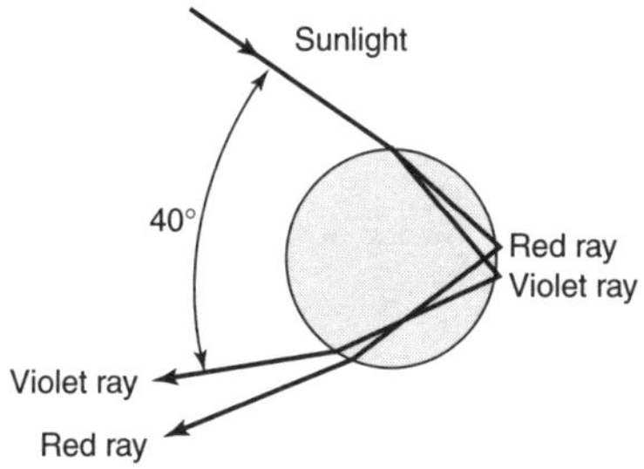

Figure 5-28 shows how the viewer sees the rainbow. The rainbow is in the shape of an arc because the circle of drops at the angle of about 40 degrees is in existence only above the ground. It is possible to see a circular rainbow from an airplane in the correct position relative to the sunlight and raindrops.

Figure 5-28 Formation of a rainbow due to total internal reflection and dispersion.

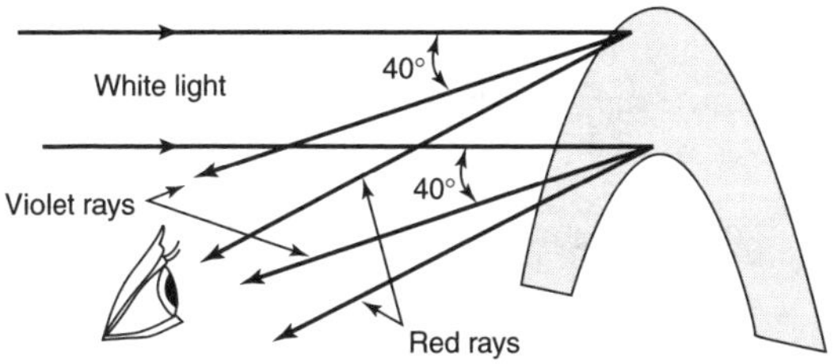

Wave Optics

To explain some phenomena, such as interference and diffraction of light, it is necessary to go beyond geometrical optics.

Huygens' principle

As mentioned earlier, Huygens considered light to be a wave. He envisioned a wave crest advancing by imagining each point along the wave crest to be source point for small, circular, expanding wavelets, which expand with the speed of the wave. The surface tangent to these wavelets determines the contour of the advancing wave. Figure 5-29 illustrates Huygens' construction for a plane wave (a) and for a spherical wave (b).

Figure 5-29 Huygens' principle for (a) a plane wave and (b) a spherical wave.

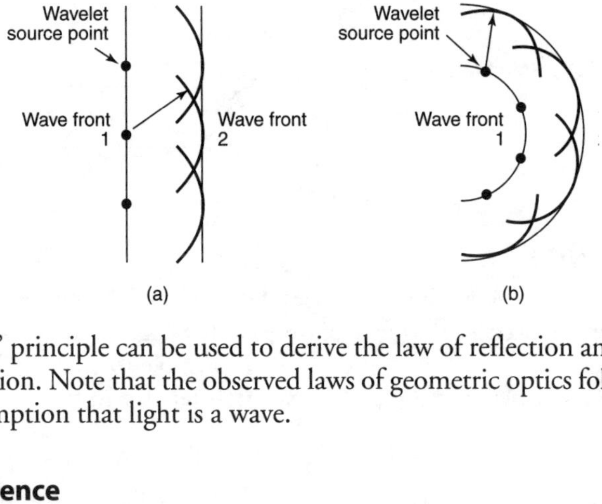

Huygens' principle can be used to derive the law of reflection and the law of refraction. Note that the observed laws of geometric optics follow from the assumption that light is a wave.

Interference

Because light is a wave, the superposition principle discussed in the wave motion section of Chapter 2 is valid to determine the constructive and destructive interferences for light waves. Interference in light waves is not easy to observe because the wavelengths are so short. For constructive interference, two waves must have the two contributing crests and the two troughs arriving at the same time. For destructive interference, a crest from one wave and a trough from the other must arrive at a given point at the same time.

Young's experiment

Thomas Young first demonstrated interference from light waves with a double slit. The schematic diagram for this experiment is shown in Figure 5-30.

Figure 5-30 Schematic diagram of Young's experiment, demonstrating interference between light waves.

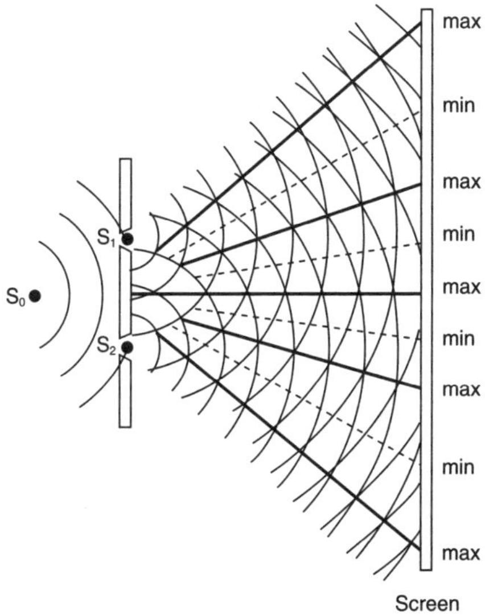

The single light source is located at S_o, and the light goes through two very narrow openings at S_1 and S_2. (A single light source is necessary because the light waves must have identical frequency and phase. The light beam is also considered to be of one colour.) Each of the slits act as a source for circular expanding waves. The points of intersection of two crests, one from each slit, are points of constructive interference. The point of intersection of a crest from one slit and a trough from the other slit is a point of destructive interference. Therefore, the interference pattern called fringes, consisting of alternating light and dark bars, will be seen on the screen.

To better understand how these points are formed, Figure 5-31 illustrates the rays coming through two slits that are directed to the point P on the screen.

Figure 5-31 The paths of two waves from the slits to the point *P*.

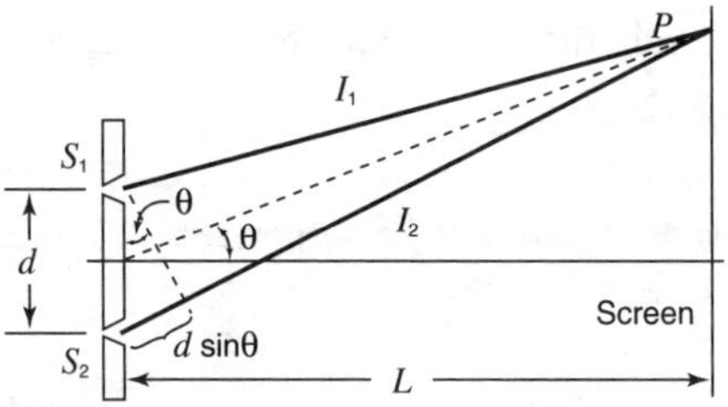

The difference in path length of the two rays is given by $d \sin \theta = l_2 - l_1$. If the path difference is a whole number of wavelengths, then constructive interference takes place. If the paths differ by a half number of wave lengths, destructive interference occurs. Using n to represent any integer, the two cases may be written

maximum brightness if $n\lambda = d \sin \theta$

minimum brightness if $\left(n + \frac{1}{2}\right)\lambda = d \sin \theta$

where λ is the wavelength and *d* is the distance between the two slits. ***Note:*** This figure is not to scale: The distance to the screen *(L)* is much greater than the distance between the slits *(d)*.

Diffraction

Young's double-slit experiment shows that light spreads out in wavefronts that can interfere with each other. **Diffraction** is the effect of a wave spreading as it passes through an opening or goes around an object. The diffraction of sound is quite obvious. It is not at all remarkable to hear sound through an open door or even around corners. In contrast, diffraction is quite difficult to observe with light. The difference is that sound waves are long while light waves are extremely short because differentiation is proportional to wave length; it is not easy to observe the bending of light when it passes through a small aperture or goes around a sharp edge.

A single slit yields an interference pattern due to diffraction and interference. Imagine that the slit is wide enough to allow a number of wavelets. Figure 5-32 shows the wave-ray diagram used to analyze the single slit.

Figure 5-32 Diffraction of light through a single slit.

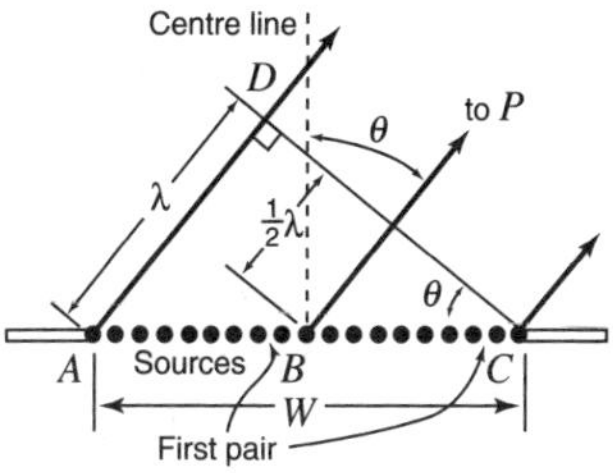

The rays from A and B interfere at P on a distant screen. As shown, AP exceeds BP by half a wavelength; therefore, the represented waves destructively interfere. Also for every wave originating between A and B, there is another point between B and C with a wavelet that will destructively interfere. The wavelets cancel in pairs; thus, point P is a minimum or dark point on the screen.

The triangle ACD is nearly a right triangle if P is quite distant. Applying the definition for sine to the figure yields

$$\sin\theta = \frac{\lambda}{w}$$

where λ is the wavelength and w is the slit width. Whenever the path difference between AP and CP is a whole number of wavelengths, a dark fringe will be produced on the screen because the wavelets can be seen to completely cancel in pairs.

Figure 5-33 illustrates the light rays traveling to another point on the screen.

Figure 5-33 Diffraction of light through a single slit.

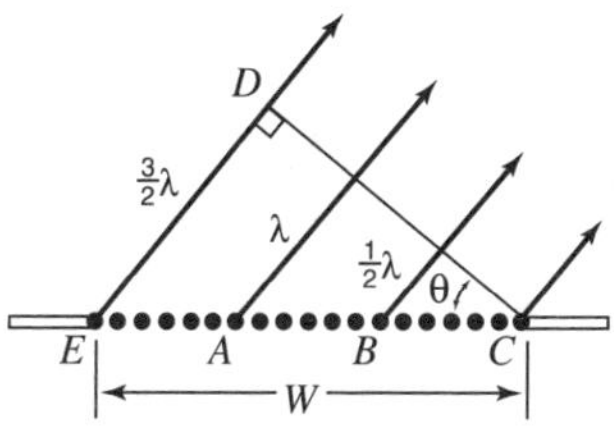

In this case,

$$\sin\theta = \frac{(3\lambda/2)}{w} = \frac{3\lambda}{2w}$$

The region of wavelets is divided into three. Again, the waves through two regions cancel in pairs, but now the waves from one region constructively interfere to produce a bright point on the screen. This is partial reinforcement. The positions of the light and dark fringes formed by a single slit are summarized in the intensity versus angle sketch shown in Figure 5-34. The centre region of the pattern will be the brightest band because the wavelets completely, constructively interfere in the middle.

Figure 5-34 Position of fringes produced by single-slit diffraction.

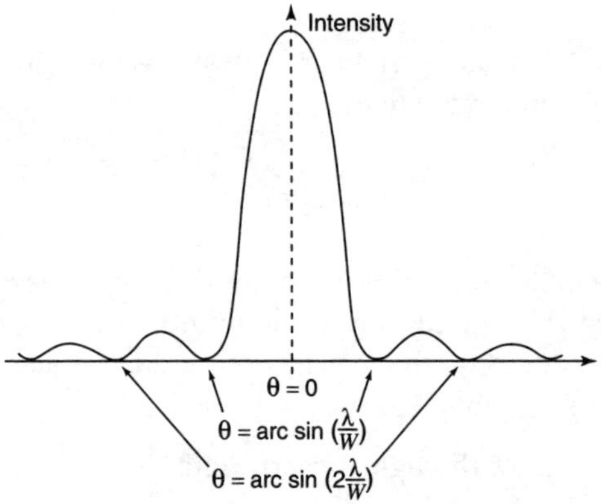

When looking through double slits, it is impossible to see only the double-slit pattern because the double-slit is really two single slits; therefore, the actual observed pattern is that of superimposed double- and single-slit patterns.

Chapter Checkout

1. True or False: The rules of geometric optics can be applied to explain diffraction of light through a single slit.
2. A fish swims 10 cm below the surface of a pool. Meanwhile, a person views it from above the pool's surface, at an angle of 30° from the normal to the pool. What angle must the light leaving the fish make with the surface of the water? Assume $n = 1.33$ for water.
3. Fibre optic cable consists of two concentric glass fibres with different indices of refraction. Assume the outer layer has $n = 1.55$ and the inner layer has $n = 1.50$. What angles of light incident on the end of the inner fibre will be propagated down the fibre by total internal reflection?
4. Suppose a light source emits at two wavelengths only: 400 nm and 600 nm. The light is incident on a single-slit diffraction apparatus with a slit width of 1600 nm. What is the angular separation between the first brightness maximums (outside of the central maximum) of the different colours?

Answers: 1. False **2.** 22.1° **3.** A cone of total angular width 30° **4.** 12.2°

Chapter 6
MODERN PHYSICS

CHAPTER CHECK-IN

- ❑ Learning about Einstein's theories of special and general relativity
- ❑ Introducing the fundamental concepts of quantum mechanics
- ❑ Applying basic quantum mechanical concepts to the atom
- ❑ Applying basic quantum mechanical concepts to the nucleus, to learn about radioactivity, fission, and fusion

The physics discussed in previous chapters was known mainly before 1900. Newton's laws of motion, Maxwell's equations, the laws of thermodynamics, and kinetic theory were successful in explaining many common phenomena. Then scientists began to look beyond the everyday world, at the worlds of high velocities, the worlds of distant stellar objects, and the worlds of atomic and subatomic particles. Modern physics refers to new developments and theories researched in the twentieth century.

Relativity

Albert Einstein devised the special theory of relativity to reconcile discrepancies between the fields of electromagnetism and mechanics. **Special relativity** is the mechanics of objects moving at high velocities, near the speed of light, in contrast to Newtonian mechanics, which deals with velocities found in daily life. Einstein is quoted as saying, "Common sense is that layer of prejudice laid down prior to the age of sixteen." This remark was prompted by the realization that the laws of special relativity are contrary to our common sense because we do not travel at speeds near the speed of light.

General relativity treats gravitational fields as equivalent to acceleration. The theory relates to the physics of the stars and even to the evolution of the universe—cosmology.

Frames of reference

Newton's laws apply in any nonaccelerating frame of reference, called an **inertial frame,** that is, one with constant velocity. According to Newtonian physics, if the laws of mechanics are valid in one inertial frame of reference, they must be valid in any inertial frame of reference. For example, if the laws of mechanics are valid in an experiment on a train moving with a constant velocity, they must also be valid at the railroad station. Imagine that a person on the train throws a ball straight up. A person observing this action from the railroad platform will insist that the ball had a parabolic path due to the forward motion of the train. Each, however, will agree that the ball obeys the law of mechanics. Furthermore, the person at the station may say that he is at rest while the train moves past in one direction; the person on the train may say that she is at rest and that the station is moving past in the opposite direction. As long as the frame of reference is nonaccelerating, there is no way to prove that any given frame of reference is absolutely at rest, nor is there a preferred frame of reference.

Michelson-Morley experiment

The statement that the laws of mechanics are valid in all frames of reference led to a contradiction when applied to Maxwell's electromagnetic equations. Maxwell predicted that the speed of light always propagated through a vacuum with a unique speed c. Classical mechanics could not explain why light should always travel with the same speed, regardless of the frame of reference.

In the nineteenth century, scientists believed that all waves must travel on some medium; thus, light was said to travel on **ether,** sometimes called **luminous ether** (not at all the same thing as the chemical compound). The ether had very unusual properties: Scientists said it had to exist in glass, other transparent materials, and even a vacuum because light travels through them; it had to be exceptionally thin because there was no discernible drag on planets in orbit; and it had to be undetectable and rigid in order to vibrate at the high frequencies of light waves. If light traveled with a constant speed with respect to a specific frame of reference, it was thought it might be that of the ether—the absolute frame.

Albert A. Michelson (1852–1931) and Edward W. Morley (1838–1923) used an interferometer to investigate the ether. A beam of light was split into two paths. One path was parallel to the direction of the motion of the earth, and the other path was perpendicular to the first. The earth's moving through the ether would be equivalent to an ether wind blowing in the opposite direction, which was parallel to the interferometer. The interference of the recombination of the two beams at the telescope gave an interference pattern of fringes. Then the interferometer was rotated through 90 degrees to see if the velocity of light changed with direction. The fringe shift was not observed. The null result was interpreted in two ways: (1) Ether does *not* exist; therefore, the electromagnetic wave does *not* need a medium for its propagation. (2) There is *no* preferred frame of reference—no absolute frame of reference.

The special theory of relativity

Einstein suggested that absolute motion has no meaning, that all motion is relative. He formulated two basic postulates for the **special theory of relativity.**

- The laws of physics are the same in all inertial reference frames.
- The speed of light is the same regardless of the frame of reference of the observer.

These simple postulates led to profoundly different ways of viewing the universe.

Addition of velocities

The first law seems reasonable; the second law is counterintuitive. In our experience, velocities simply add. For example, if a person throws a ball toward you at 10 m/s while riding a bicycle towards you at 10m/s, the ball seems to come at you at 20 m/s. The expression would be $u = u' + v$, where u is the velocity with respect to the ground, u' is the velocity of the ball with respect to the bicycle, and v is the velocity of the bicycle.

The second law of special relativity leads to a different scenario for light. If a person rides toward you at 90 per cent the speed of light and turns on a flashlight, you would find the light coming toward you at c, the speed of light. Moreover, this is the same speed observed if the bicycle were stationary relative to your position.

The different results in adding velocities from experience and from experiments with light beams could be explained only by assuming that adding low velocities is a special case of a more general law for the addition of velocities. Fizeau experimentally determined the correct equation for adding velocities to be

$$u = \frac{u' + v}{1 + u'v/c^2}$$

This equation yields the following equation for the speed of a light beam from a rider coming toward you with a speed of v:

$$u = \frac{c+v}{1+\frac{cv}{c^2}} = \frac{c+v}{1+\frac{v}{c}} = \frac{c+v}{\frac{c+v}{c}} = \frac{c(c+v)}{c+v} = c$$

This equation for adding velocities gives the correct results as predicted by the second postulate of special relativity. Try other values to see that the combination of velocities less than c, and even equal to c, cannot give a combined value greater than c. Notice that if u' and v are relatively small, the denominator becomes 1; then the expression reduces to the familiar equation for the addition of velocities.

Time dilation and the Lorentz contraction

The consequence of Einstein's postulates is that the measurements of time and distance are *not* constant when compared from one inertial frame of reference to another so that the speed of light can remain a constant. In other words, there is no such thing as absolute time and absolute distance. All clocks in the universe do *not* keep time together. **Time dilation** is the effect that a moving clock runs slower than an identical stationary clock. The expression is

$$\Delta t = \frac{\Delta t_o}{\sqrt{1-\frac{v^2}{c^2}}}$$

Furthermore, not only does the clock run slower but so also does any physical process affected by the passing of time such as chemical and biological processes. Therefore, to the person in the moving frame, no change in time interval can be detected because all methods of measuring time are slowed by the same factor. All is relative.

Time dilation has been measured in a number of experiments. One experiment involved the change of time for decay of radioactive particles (muons) traveling to the surface of the earth at nearly the speed of light. The number of muons were counted at the top and at the bottom of a mountain. More muons were detected at the bottom than would be expected by classical theory. The time to decay was longer, and the value was consistent with the theory of special relativity. Accurate atomic clocks carried on airplanes and in orbiting satellites have all borne out the theory.

The **Lorentz contraction** is the effect that an observer moving with an object of a given length will find the object to be shortened compared to an observer at rest relative to this motion. (Note that this shortening occurs only in the direction of motion.) Again, all is relative. Because all distances in the direction of motion are shortened for the moving observer, there is no comparison length for the traveling observer to use to detect the change. The equation is

$$l = l_o\sqrt{1 - \frac{v^2}{c^2}}$$

To connect the concept of length contraction to time dilation, imagine that you are on a muon coming to earth at nearly the speed of light. An observer on the ground could say it is now time for the particle to decay, but you could reply that by your clock the time has not yet arrived. Alternatively, if you calculate the height of the mountain, you will find it shortened in your frame of reference. In a sense, both time and distance are doctored up so that the speed of light is always the same.

The twin paradox

Imagine one twin in a rocket ship traveling toward a star and returning at nearly the speed of light. The other twin stays home. The stay-at-home twin insists that she stayed at rest and the other twin moved. The twin in the rocket ship insists that she was at rest while the earth zoomed away from her and then returned.

The contradiction is that each can claim that the other moved and, therefore, was the one who aged less. Who did age the most? (It is not possible to have relative grey hair.)

To resolve the paradox, realize that the problem is not symmetric. When the rocket ship twin was leaving the earth, slowing to a stop in outer space, turning around to come home, and slowing to a stop at the earth's surface, she accelerated and decelerated; therefore, her frame of reference was not an inertial frame of reference. The traveling twin aged less.

Relativistic momentum

The definition of momentum and energy must be generalized to fit within special relativity. The correct expression for relativistic momentum is

$$p = \frac{m_o v}{\sqrt{1 - \frac{v^2}{c^2}}}$$

Note that when v is low, the denominator is nearly 1, so momentum approaches the familiar equation: $p = mv$. This equation can be interpreted to mean that the mass is velocity-dependent:

$$m = \frac{m_o}{\sqrt{1 - \frac{v^2}{c^2}}}$$

When v is zero, the mass equals m_o. The term m_o is called the **rest mass.**

Relativistic energy

Einstein proposed the famous **mass-energy equivalence** equation: $E = mc^2$. This energy is the sum of the kinetic and rest energies. The relationship shows that mass is a form of energy; therefore, a statement of conservation of energy must include the concept of mass. Because c is such a large value, a small mass is equivalent to an enormous energy.

The kinetic energy of a mass is

$$K.E. = m_o c^2 \left[\frac{1}{\sqrt{1 - \frac{v^2}{c^2}}} - 1 \right]$$

By using the binomial theorem to expand the expression, it can be shown that for small v this expression simplifies to the familiar

$$\frac{1}{2}(mv^2)$$

General relativity

The **general theory of relativity** is a theoretical framework applicable to any frame of reference—inertial or accelerating. In developing this theory, Einstein wanted to produce a theory of gravitation that incorporated the theory of special relativity and the equivalence principle.

To understand the equivalence principle, imagine that you are in a spaceship moving through outer space and cannot see outside the craft. After freely floating inside the craft, you start to drift to one end. Soon, you can stand on one wall. Either of two conditions could be in effect: (1) the spaceship is accelerating so that you are forced against the wall opposite the direction of acceleration, or (2) the ship has come close to a large mass with a gravitational field. The **equivalence principle** is that experiments conducted in either a uniformly accelerating frame of reference or in an inertial frame of reference with a gravitational field give the same results.

Einstein determined that the concepts of space should be reexamined. Newton's laws presuppose a Euclidean space extending in all directions, like three mutually perpendicular straight axes. Einstein proposed a **curved space time.** The motion of an object could be described in terms of the geometry of space, instead of as a response to applied forces.

For example, light travels in straight lines in Euclidean space; however, the general theory of relativity views space as distorted by mass. The light, therefore, might travel a curved path near a massive object, which is the shortest path between two points in that space. This is analogous to a curved path on the earth's surface, which is a shorter distance between two cities than a straight line between them.

To test his theories, Einstein proposed three phenomena that could be explained by the mathematical formulations of the general theory of relativity:

- The point of closest approach—the perihelion—of the elliptical orbit of Mercury advances about the sun. (Figure 6-1 illustrates an exaggerated view of this motion.)
- The deflection of starlight passing near the sun is due to the warping of space near the solar mass.
- The gravitational redshift of light is a frequency shift between two identical clocks at different heights in a gravitational field.

Figure 6-1 Advance of the perihelion of Mercury.

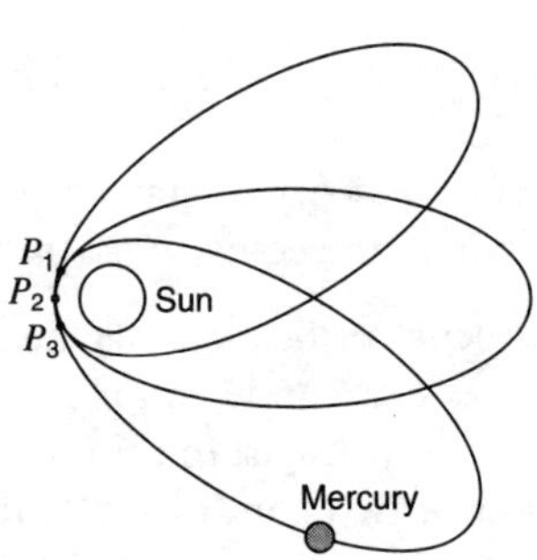

When Einstein suggested these tests, only the first phenomenon had been observed. Subsequently, scientists found and measured the other effects. All of the experimental data is consistent with the theory of general relativity.

Quantum Mechanics

Not only was classical mechanics unsuccessful in explaining motions near the speed of light, it also could not explain the behaviour of matter on the atomic level. **Quantum mechanics** is required to analyze the behaviour of molecules, atoms, and nuclei.

Blackbody radiation

A **blackbody** is an ideal thermal object that absorbs all radiation falling on it at low temperatures and is also a perfect radiator. The curves of radiation intensity versus wavelength could not be explained by classical physics. Max Planck (1858–1947) developed an equation for blackbody radiation that agreed with the data. This derivation required two assumptions:

- The vibrating molecules emitting the radiation could have only certain discrete amounts of energy, given by $E_n = nhf$, where n is called a **quantum number,** f is the frequency, and h is **Planck's constant** given by $h = 6.626 \times 10^{-34}$ joule-seconds.
- Molecules emit energy in units called **quanta,** now called **photons.** They do this by jumping from one energy state to another. The energy of the light quanta emitted by the jump between energy states is given by $E = hf$, or in terms of wavelength,

$$E = \frac{hc}{\lambda}$$

The radical nature of Planck's vision is the assumption of quantized energy states. The terms **discrete** and **quantum** referred to considering the energy as coming in packets instead of as a continuous flow; thus, the molecule will change energy states only if the amount of energy absorbed or radiated is a discrete amount of energy.

Photoelectric effect

The **photoelectric effect** is the emission of electrons from certain metals when light shines on the metallic surface. The emitted electrons are called **photoelectrons.** A number of aspects of the effect were puzzling:

- For a given metal, the light has to be of at least a minimum frequency, called the **cutoff frequency** or the **threshold frequency.**
- The kinetic energy of the photoelectrons is *not* dependent upon the intensity of the light causing the photoelectric effect.
- The maximum kinetic energy of the photoelectrons increases with increasing light frequency.
- Electrons are emitted almost immediately from the surface; that is, no build-up of energy flowing into the molecules is necessary. Figure 6-2 shows a sketch of the maximum kinetic energy versus the frequency of light incident upon the metal.

Figure 6-2 The photoelectric effect.

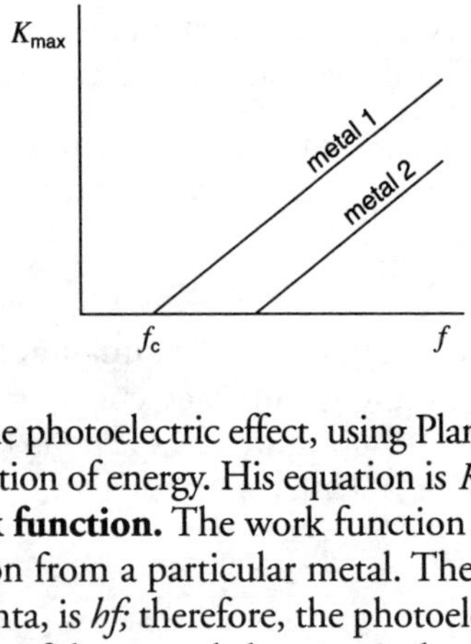

Einstein explained the photoelectric effect, using Planck's quantum hypothesis and the conservation of energy. His equation is $K.E._{max} = hf - Q$, where Q is called the **work function.** The work function is the energy required to release the electron from a particular metal. The energy of the incoming photons, or quanta, is hf; therefore, the photoelectric equation simply states that the energy of the ejected electrons is the difference between the energy absorbed from the quanta of light and the energy required to escape from the material.

The unexplained observations described above can be illuminated by the following arguments.

- A certain threshold frequency of the incident light is necessary to provide a quanta of sufficient energy to eject an electron.
- As the intensity of the incident light increases, so will the number of emitted electrons, but their kinetic energies will not increase.

- As the frequency increases, so does the energy of the photons so that the ejected electrons will have greater kinetic energy.
- A low-light intensity indicates few quanta, but if those are of sufficient energy, some electrons will be emitted immediately.

Compton scattering

Additional evidence for the quantized nature of electromagnetic waves came from the Compton effect, named for Arthur Compton (1892–1962). **Compton scattering** involves the scattering of a high-energy, e-ray photon and an electron. The scattered photon has less energy than the original photon, which can be seen as a change in wavelength. Compton explained this by assuming that the photon behaves like a particle when interacting with the electron. The conservation of momentum and energy used for elastic collisions of billiard balls could mathematically explain the experimental observations. The scattering effect is dependent upon the angle but not the wavelength. The small shift in wavelength would be too difficult to detect with less energetic photons, such as light photons.

Particle-wave duality

The photoelectric effect and the Compton effect again point to the duality of the nature of electromagnetic radiation. The models of light as a wave and also as a particle complement each other. When the photons of electromagnetic radiation are of relatively high energy, the wavelengths are short. Then the photon acts more like a particle than a wave. For example, the Compton photons were high-energy x-rays. When the photons of electromagnetic radiation have relatively low energy, the wavelengths are long. Radio waves are an example of less energetic photons that act more like waves than particles.

De Broglie waves

Louis de Broglie (1892–1987) postulated that because photons have both wave and particle characteristics, perhaps particles also have wave characteristics. From the energy of the photon,

$$E = hf = \frac{hc}{\lambda}$$

the momentum of a photon can be derived:

$$p = \frac{E}{c} = \frac{hc}{c\lambda} = \frac{h}{\lambda}$$

De Broglie hypothesized that material particles with momentum p should have a wave nature and a corresponding wavelength given by his equation:

$$\lambda = \frac{h}{p} = \frac{h}{mv}$$

Note that the de Broglie wavelength is directly proportional to h, which is a constant to the –34 power. With the relatively large masses and velocities of ordinary life, the de Broglie wavelengths are so small that they are virtually undetectable.

The Heisenberg uncertainty principle

The **Heisenberg uncertainty principle**, formulated by Werner Heisenberg (1901–1976), states that it is impossible to simultaneously measure a particle's position and velocity exactly. Specifically, the uncertainty in the measurements are given by

$$(\Delta x)(\Delta p) \geq \frac{h}{2\pi}$$

Another form of the expression refers to the uncertainty in measurements of energy and time

$$(\Delta E)(\Delta t) \geq \frac{h}{2\pi}$$

In principle, it is possible to make exact measurements in classical physics; however, even in principle making exact measurements is not possible in quantum mechanics. Consider finding the exact position of a charged particle that produces a spot of light when hitting a phosphor. The exact position is known, but information about the particle's momentum has changed. Or consider viewing an object under a microscope. In order to see the object, some photons must reflect off it to the eye of a viewer. These incident photons will cause uncertainties in the measurement. In other words, the very act of the measurement procedure in quantum mechanics introduces uncertainty into the data collected.

Atomic Structure

Early in the study of atomic structure, Sir Joseph Thomson and Ernest Rutherford produced competing models of the atom. Thomson proposed the "plum-pudding" model. In this model, the negative charges—**electrons** (the plums)—were surrounded by the positive charges (the pudding), filling the volume of the atom.

To demonstrate his model, Rutherford bombarded thin metal foils with a beam of positively charged particles. Most of the particles went through with little effect, but occasionally one was deflected through a large angle. By way of explanation, Rutherford proposed a planetary model of the atom with the negatively charged electrons orbiting a central concentration of positive charge—the **nucleus.**

Two main difficulties occur in the planetary model of the atom. First, any object moving in a circle is accelerating. According to classical mechanics, an accelerating charge radiates energy, producing electromagnetic waves; therefore, the orbiting electron should radiate energy and fall quickly into the nucleus. Because matter does exist, this obviously does not happen. Second, an atom emits only certain electromagnetic radiation, *not* a continuum of all frequencies.

Atomic spectra

When a gas is excited by a spark, light of a particular colour is produced. For example, neon gas produces a red-orange colour. When this light is spread through a prism, a series of bright lines of specific wave lengths are observed, called a **line spectrum.** The line spectra are characteristics of a certain gas, rather like its atomic fingerprint. For example, the visible line spectrum of hydrogen consists primarily of wavelengths of approximately 656 nm, 486 nm, 434 nm, and 410 nm. Although it was not understood at the time why it worked, the following equation was found to describe the series of lines, called the **Balmer series:**

$$\frac{1}{\lambda} = R\left(\frac{1}{2^2} - \frac{1}{n^2}\right)$$

where n is an integer of 3, 4, 5, . . . and R is a constant, now called the **Rydberg constant:** $R = 1.0973732 \times 10^7\, m^{-1}$.

In addition to emitting light of certain wavelengths, an element can also absorb light of specific wavelengths. The light from a light bulb or the sun forms a **continuous spectrum**—the colours of the rainbow. When a continuous spectrum light is passed through a cool gas, the gas absorbs the same wavelengths that it emits when excited, which appear as a set of black lines in the continuous spectrum. The black lines are called the **absorption spectrum.**

The Bohr atom

Niels Bohr (1852–1962) combined classical mechanics and some revolutionary postulates to formulate a model of the hydrogen atom that hoped to circumvent some of the difficulties of classical physics and still explain atomic spectra. The following are his basic postulates.

- Postulate 1: The electron moves in only certain permitted circular orbits—quantized states—around the positive nucleus under the influence of the Coulomb force.
- Postulate 2: The electron does not emit energy when it is in one of the allowed orbits called a **stationary state.**
- Postulate 3: When the electron jumps from one permitted state to another, the energy is given off as a particular photon with energy equal to the difference in the energies of the initial and final states:

$$hf = E_i - E_f.$$

Planck's concepts of quantization can be seen in Bohr's postulates 1 and 3. When the electron is in a stationary state, Bohr assumed that Newton's laws, Coulomb's law, and conservation of energy were valid. Bohr showed that the angular momentum of an electron with mass *m* traveling with speed *v* about a circular orbit of radius *r* is quantized as

$$L = mvr = \frac{nh}{2\pi}$$

where *n* is an integer and *h* is Planck's constant.

Also, he derived an expression for the radius of hydrogen from the electrostatic force (Coulomb's law) set equal to the centripetal force:

$$r_n = \frac{\varepsilon_o n^2 h^2}{\pi m e^2}$$

When $n = 1$, the radius is called the **Bohr radius,** which is the smallest orbit of hydrogen.

To find an expression for the total energy of the electron orbiting the atom, use the classical formula for total energy, then substitute r_n from above and *v* from the angular momentum to get the following:

$$E = K.E. + P.E. = \frac{1}{2}mv^2 - \frac{e^2}{4\pi\varepsilon_o r_n}$$

When $n = 1$, the lowest energy state of the atom is called the **ground state.** The value of the ground state of hydrogen is –13.6 electron volts, which is in excellent agreement with the experimentally observed hydrogen

ionization energy—the energy necessary to remove an electron in the ground state from an atom.

Combining this result with the equation in Bohr's postulate 2 yields

$$hf = E_f - E_i = \frac{-me^4}{8\varepsilon_o^2 h^2 n_f^2} - \frac{-me^4}{8\varepsilon_o^2 h^2 n_i^2}$$

Because $c = f\lambda$, the equation becomes

$$\frac{1}{\lambda} = \frac{me^4}{8\varepsilon_o^2 h^3 c}\left(\frac{1}{n_i^2} - \frac{1}{n_f^2}\right)$$

From the preceding equation, the Rydberg constant may be calculated.

$$R = \frac{me^4}{8\varepsilon_0^2 h^3 c}$$

All of these constants are known, and the theoretical value for the Rydberg constant is the same as this derived R. This demonstrated agreement is remarkable, and it validated Bohr's postulates.

Energy levels

The Balmer series, found experimentally, can be explained by the Bohr model of the atom in the following way. Figure 6-3 is a diagram of the energy transitions possible for hydrogen.

Figure 6-3 Energy transitions of a hydrogen atom, with the spectral series.

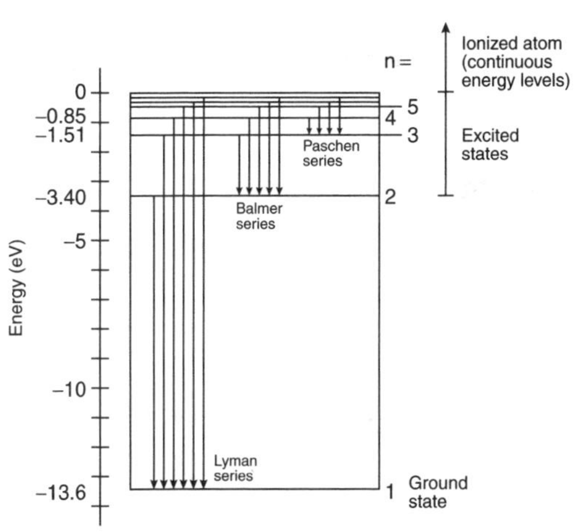

For the Balmer series, the hydrogen electron jumps from an initial excited state ($n = 3,4,5, \ldots$) to a final state at the $n = 2$ level. In so doing, it emits a photon with energy equal to the energy difference of the initial and final states. Other series indicated on Figure 6-3 illustrate the other series of lines found by Theodore Lyman and Louis Paschen. This type of diagram is called an **energy level diagram** because it illustrates the discrete, allowed energy levels and the permissible transitions for the orbiting electron.

De Broglie waves and the hydrogen atom

The next task was to suggest why only certain discrete energy levels are possible. De Broglie assumed that an orbit would be stable only if it contained a whole number or multiples of a whole number of electron de Broglie waves. Figure 6-4 shows a representation of a standing circular wave of three wave lengths.

Figure 6-4 DeBroglie waves of an electron orbiting with $n = 3$.

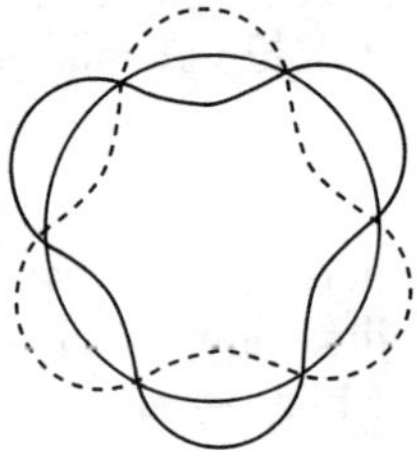

The depicted orbit would be the permissible orbit with the quantum number of 3, that is, $n = 3$. This visual way of understanding quantization shows that the wave nature of matter is basic to a model of the atom. More complicated formulations of quantum physics that were developed later have built on these concepts.

Nuclear Physics

Nuclear physics, as the name implies, deals with the model and mechanics of the nucleus.

Nucleus structure

Building upon the understanding of a central positive charge—called a **proton**—in the nucleus of hydrogen, it seemed reasonable to assume that other atoms also had nuclei with protons.

Sir James Chadwick demonstrated the existence of a neutral particle—called a **neutron**—that has essentially the same mass as the proton. The nucleus is made of protons and neutrons that, collectively, are called **nucleons.** According to the modern model of the nucleus, the **atomic number** *(Z)* is the number of protons in the nucleus, and the **atomic mass** *(A)* is the number of nucleons in the nucleus. (The number of electrons is equal to the number of protons in an electrically neutral atom, and so the number of orbiting electrons is also given by the value of *Z*.) Nuclei with the same number of protons but differing number of neutrons are called **isotopes.** The chemical properties of an element are determined by the outer electrons (equal to the number of protons); therefore, isotopes are identical in chemical nature but differ in mass. The symbol for an element *(X)* is ${}^{A}_{Z}X$; for example, ${}^{9}_{4}\text{Be}$ is beryllium with four protons and five neutrons.

Binding energy

When the masses of the constituent particles of a nucleus are added together, the sum is less than the nucleus itself. For example, a deuteron is an isotope of hydrogen with one proton and one neutron in the nucleus. The following below adds these particles in **atomic mass units**—abbreviated here as amu—where 1 amu is 1/12 of a carbon atom with 12 nucleons.

$$\begin{aligned}\text{The mass of 1 proton} &= 1.007825 \text{ amu}\\ \text{The mass of 1 neutron} &= \underline{1.008665} \text{ amu}\\ & 2.016490 \text{ amu}\end{aligned}$$

The observed mass of the deuteron is 2.014102 amu, which is .002388 less than the sum. Using the mass-equivalence equation, $E = mc^2$, 1 amu corresponds to approximately 931 MeV. Thus, the mass difference is (0.002388 amu)(931 MeV/amu) = 2.224 MeV. This quantity is called **binding energy.** The binding energy is the difference between the mass energy of the nucleus and its constituent particles. To separate the nucleus into a proton and neutron, energy equal to the binding energy must be added to the system.

Radioactivity

Some nuclei are unstable and spontaneously emit radiation, which is called **radioactivity.** The radiation is of three types:

- Alpha decay, in which the emitted particles are helium nuclei of 2 protons and 2 neutrons

- Beta decay, in which the emitted particles are electrons
- Gamma decay, in which high energy photons are emitted

The original nucleus is called the **parent nucleus,** and the nucleus remaining after the decay is called the **daughter nucleus.** The process of one element changing into another through radioactivity is called **transmutation.**

If a nucleus emits an alpha particle, it loses two protons and two neutrons; therefore, the daughter nucleus has an atomic mass of 4 less and an atomic number of 2 less than the parent nucleus. An example of alpha decay of uranium is ${}^{238}_{92}\text{U} \rightarrow {}^{234}_{90}\text{Th} + {}^{4}_{2}\text{He}$.

If a nucleus emits a beta particle, it loses an electron. Since the mass of the electron is so small compared to that of a proton and a neutron, the atomic mass of the parent nucleus is the same as the daughter nucleus. The atomic number of the daughter nucleus is one greater than that of the parent nucleus. An example of beta decay of bismuth is ${}^{212}_{83}\text{Bi} \rightarrow {}^{212}_{84}\text{P}_{\text{o}} + {}^{0}_{-1}e$.

Frequently the daughter nucleus is left in an excited state after either alpha or beta decay. Then the nucleus can give up excess energy by emission of gamma radiation. The following example shows a typical situation where gamma decay occurs: ${}^{12}_{5}\text{B} \rightarrow {}^{12}_{6}\text{C}^{*} + {}^{0}_{-1}e$; then, ${}^{12}_{6}\text{C}^{*} \rightarrow {}^{12}_{6}\text{C} + \gamma$, where the asterisks indicate an excited nucleus.

The rules for radioactive decay are based on conservation laws. Examination of the preceding examples reveals that the number of nucleons and the electric charge are conserved; that is, the total on one side of the equation equals the total on the other side of the equation. Other conservation laws that must be observed are those of energy, momentum, and angular momentum.

Half-life

The **decay rate** *(R)* or the **activity** of a sample of radioactive material is defined as the number of decays per second, given by $R = -\lambda N$, where N is the number of radioactive nuclei at some instant and λ is the **decay constant.**

The **half-life** *(T)* is defined as the time required for half of a given number of radioactive nuclei to decay. It is different for each type of radioactive element:

$$T = \frac{0.693}{\lambda}$$

The general decay curve for a radioactive sample relating the number of nuclei present at a given time to the original number of nuclei is exponential. The expression is $N = N_o e^{-\lambda t}$, where N is the original number of nuclei, N_o is the number of nuclei at time t, and e is the base of the natural logarithm.

Nuclear reactions

Nuclear fission occurs when a heavy nucleus splits into two nearly equal size nuclei. The reaction for uranium 235 is ${}^{1}_{0}n + {}^{235}_{92}\text{U} \rightarrow {}^{141}_{56}\text{Ba} + {}^{92}_{36}\text{Kr} + 3\,{}^{1}_{0}n$. The total rest mass of the products is less than the original rest mass of the original uranium by 220 MeV. This is an enormous amount of energy compared to energy releases in chemical processes and when considering that a relatively modest piece of uranium has so very many nuclei. **Nuclear fusion** occurs when light nuclei are combined to form a heavier nucleus. The sun is powered by nuclear fusion.

The binding energy is related to stability. When the mass energy of the parent nucleus is greater than the total mass energy of the decay products, spontaneous decay will take place. If the decay products have a greater total mass energy than the parent nucleus, additional energy is necessary for the reaction to occur. Energy is released when light nuclei combine (fusion) and when heavy nuclei split (fission).

Chapter Checkout

1. The mother ship is traveling through space at a constant velocity of $0.9c$, relative to an observer on a nearby moon. A shuttle is launched from the mother ship at $0.5c$ relative to the mother ship. What speed does the shuttle have with respect to the observer on the moon?
2. A muon travels through the upper atmosphere with a velocity of $0.85c$ relative to the ground. The lifetime of a muon *at rest* (or, in its own reference frame) is 1 μs. The muon is at the top a mountain which is 10,000 m tall, measured in its rest frame. **(a)** From the muon's perspective, how high is the mountain? How long would it take the muon to travel the length of the mountain? **(b)** For an observer on the mountain, what is the lifetime of the muon? How far does the observer see the muon travel in that time? **(c)** How would you reconcile the experiences of the muon and the observer? Does the muon reach the ground?

3. A spaceship captain orders his crew to accelerate their ship to the speed of light. **(a)** How much energy is required to increase the speed of the ship from $0.1c$ to $0.2c$, if the ship has a rest mass of 10^5 kg? **(b)** How much energy is required to accelerate between $0.9c$ and $0.95c$? **(c)** From $0.98c$ to $0.99c$? What would you tell the captain?

4. **(a)** Calculate your wavelength and momentum, if you have a mass of 80 kg and are riding a skateboard at 1 m/s. **(b)** Calculate the momentum of a photon, assuming it has a wavelength of 650 nm.

5. **(a)** Calculate the value of the Bohr radius. **(b)** Consider an ionized atom of helium, which would have one electron orbiting a nucleus with two protons (and two neutrons). What is the radius of the $n = 1$ state in this case? You will need to repeat Bohr's calculation of the centripetal force due to Coulomb's law.

Answers: 1. $0.965c$ **2a.** 5268 m; 20 μs **b.** 1.9 μs; 484.5 m **c.** The muon doesn't reach the ground **3a.** 1.4×10^{20} J **b.** 8.1×10^{21} J **c.** 1.9×10^{22} J **4a.** $\lambda = 8.25 \times 10^{-36}$ m; p = 80 kg m/s **b.** 10^{-27} kg m/s **5a.** 0.53 A **b.** 0.265 A

REVIEW

Use this Review to practice what you've learned in this book. After you work through the review questions, you're well on your way to achieving your goal of understanding physics.

Chapter 1

1. True or False: Doubling the distance between two objects causes the gravitational force between them to fall by 50%.
2. For each of the following physics quantities, give the appropriate SI units. In addition, give each unit in terms of the fundamental units kg, m, and s.
   a. Force
   b. Torque
   c. Momentum
   d. Energy
   e. Angular momentum
   f. Density
   g. Pressure
3. A girl swings a ball attached to a string above her head. If the string is 1.3 m long and circles the girl in 2.4 s (assume the mass of the string is not important):
   a. What is the linear velocity of the ball?
   b. If the ball has a mass of 135 g, how much force must she supply?
4. Sally and John want to balance each other on a seesaw so that the board is parallel to the ground and they are motionless. The seesaw has a total length of 5 m, with the pivot in the centre. John has a mass of 50 kg, and Sally has a mass of 35 kg. If Sally sits on the very end of the board, where must John sit?
5. Suppose you have just crash-landed on an unexplored planet. Unfortunately, most of your equipment was destroyed in the crash. Describe a simple way that you could measure the acceleration due to gravity on the new planet without using a scale.
6. Window ornaments are often hung with a suction cup. Why does the cup hold onto the window? What force prevents it from falling off?

7. Write an equation describing the relationship between **A, B, C, D,** and **E** in the following figure.

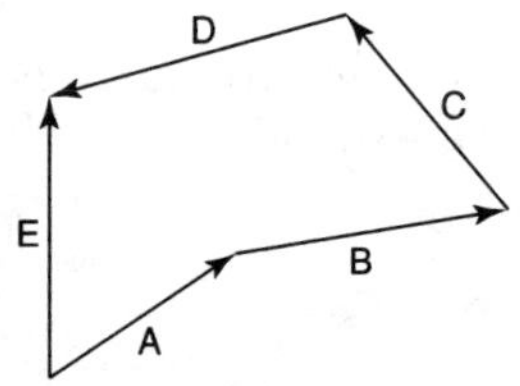

8. What is the total work done on the object feeling the force plotted here?

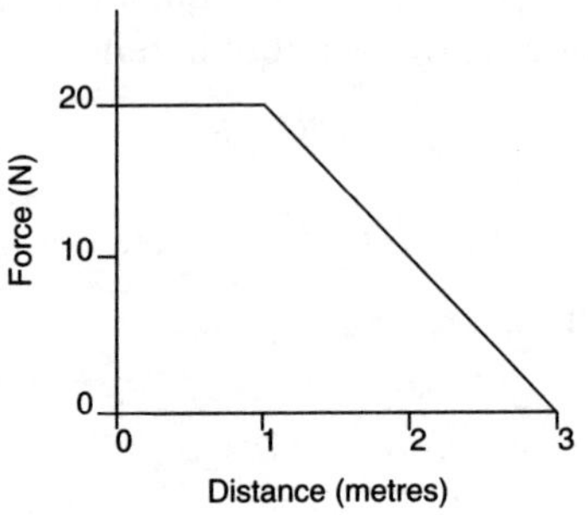

9. Find the acceleration of the object whose motion is plotted here.

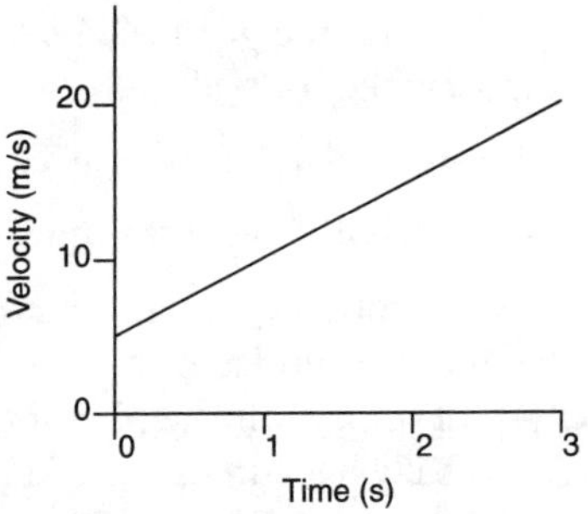

Chapter 2

10. The wave pictured here is a square wave. What characteristic is labeled *x*?

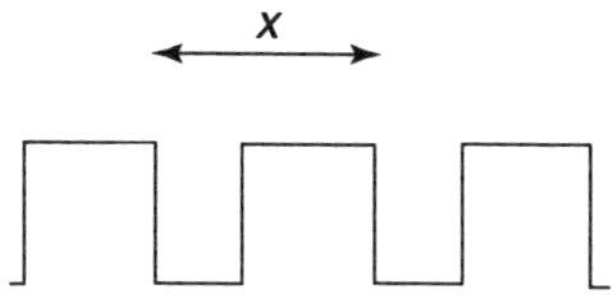

11. For each of the following physics quantities, give the appropriate units. In addition, give each unit in terms of the fundamental units kg, m, and s.

a. Frequency

b. Intensity

c. Velocity

12. A string is tied off at both ends to fixed bolts, and has total length L. What are the three largest wavelengths possible for standing waves in the string? Remember that the end points must be fixed.

Chapter 3

13. True or False: Work is done on a gas when the pressure in a container is increased without changing the volume of the container.

14. An aluminum ($\alpha = 23 \times 10^{-6}/°C$) and a steel bar ($\alpha = 11 \times 10^{-6}/°C$) meet end to end, with no gap, at a temperature of 22°C. If they each have, at this temperature, a length of 35 cm, how big will the gap be at 0°C? Assume that the opposite ends of the bars are absolutely fixed in place.

15. In a physics laboratory, an ultrahigh vacuum chamber has an air pressure of 10^{-13} Pa. At 22°C, how many molecules of gas is this for each cubic metre?

16. Why does a windy day feel colder than a day at the same temperature, but with no wind?

17. Calculate the work done in each of the thermodynamic processes whose P-V graph is shown here.

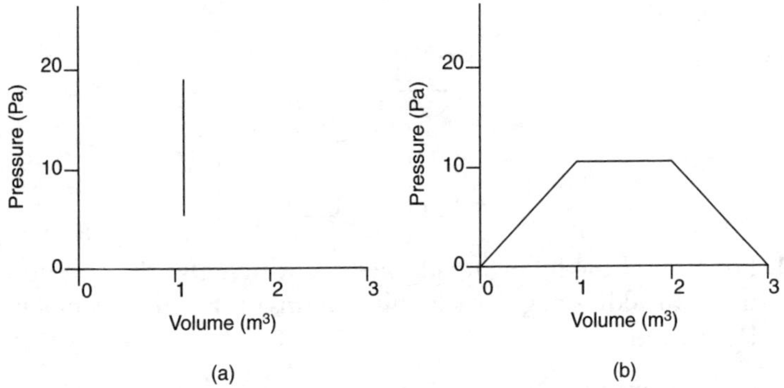

Chapter 4

18. For each of the following physics quantities, give the appropriate units. In addition, give each unit in terms of the units kg, m, s, and C.

a. Magnetic field magnitude

b. Electrostatic potential

c. Capacitance

d. Resistance

e. Inductance

19. Two concentric spherical shells carry charges of 1.5 μC and –2 μC on the inner and outer sphere, respectively. The inner sphere has radius 1 cm and the outer has radius 5 cm. Find the magnitude and direction of the electric field at the following distances from the centre of the spheres.

a. 0.5 cm

b. 2.5 cm

c. 10 cm

20. A large flat plate is maintained at an electrostatic potential of +10 V.

a. What is the magnitude and direction of the electric field 1 m from the plate's surface? (The electron mass is 9.11×10^{-31} kg.)

b. How much energy is gained by an electron moving from infinity (V=0) to the surface of the plate?

c. According to classical mechanics, how fast will the electron be traveling when it impacts the plate?

21. Three capacitors are wired in series with each other. They have capacitances 1 F, 2 F, and 3 F.

a. Find their total effective capacitance.

b. Suppose a fourth capacitor with C = 4 F is added, in parallel with the combination of the other three. What is the final effective capacitance?

c. What is the reactance of the final circuit (all four capacitors) if an ac current is passed with a frequency of 100 Hz?

22. Calculate the force per metre of length on a long straight wire carrying 1 A of current, due to a parallel wire carrying the same magnitude of current in the opposite direction. The distance between the wires is 1 cm. Don't forget to indicate the direction of the force.

23. You have learned that electric currents—that is, moving electric charges—are responsible for creating magnetic fields. How does a bar magnet work, then? Where are the currents?

24. Electrons are "negatively" charged and protons are "positively" charged. Would the laws of physics change if these definitions were reversed?

25. Examine the following circuit.

a. Are R and C_1 in series or parallel with each other?

b. Is C_1 in parallel or series with C_2?

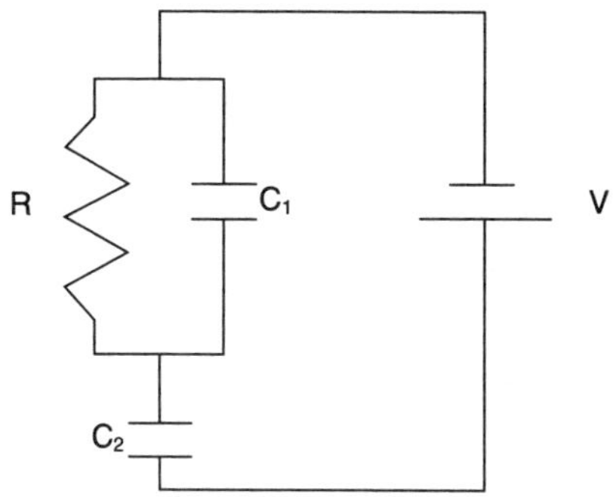

26. For the electrical current shown here, use the right hand rule to determine the direction of the magnetic field at the centre of the loop.

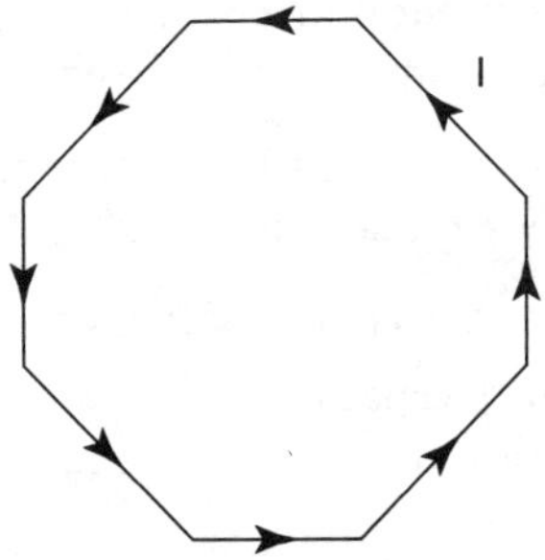

Chapter 5

27. True or False: Visible light is the only type of electromagnetic radiation.

28. The index of refraction of most glass depends slightly on the wavelength of the light passing through. Suppose that for a certain piece of glass, $n = 1.47$ for $\lambda = 400$ nm and $n = 1.46$ for $\lambda = 550$ nm. One ray of light at each wavelength is incident on the surface of a piece of this glass at an angle of 45°.

a. Give the angle from the normal for the refracted rays.

b. If the glass is 10 cm thick, how far apart will the rays be upon reaching the other side of the glass?

Chapter 6

29. True or False: It is impossible to measure the position and velocity of a proton exactly.

30. An electron is at rest. Keeping in mind that the rest mass of the electron is 9.11×10^{-31} kg, answer these questions:

a. What is the energy equivalence of its rest mass?

b. How much energy is required to accelerate it to a velocity of $0.99c$?

c. At this speed, what is the ratio of its apparent mass to its rest mass?

31. An electron in a hydrogen atom is in the third excited state ($n = 3$).

a. What is its average distance from the proton?

b. What wavelength of photon will cause it to jump to the $n = 5$ state?

c. What is the energy of that photon?

d. How much did the angular momentum of the electron change in its transition from $n = 3$ to $n = 5$?

32. An asteroid hurls past your spaceship with relative velocity $0.9c$. Your sensors indicate that the asteroid is 53 km long in the direction parallel to your motion and 76 km long in the direction perpendicular to your motion and is approximately a rectangular box shape. In the asteroid's rest frame, what are its dimensions?

33. In some science fiction stories, spaceships are propelled by sunlight. Explain why this is possible and try to estimate the size of the "sail" necessary to push a large spacecraft, such that it obtains relativistic speeds (for example, $v \sim 0.5c$).

Answers: 1. False **2a.** N, kg-m/s^2 **b.** N-m, kg-m^2/s^2 **c.** kg-m/s **d.** J, kg-m^2/s^2 **e.** J-s, kg-m^2/s **f.** kg/m^3 **g.** Pa, kg/m-s^2 **3a.** 3.4 m/s **b.** 1.2 N **4.** 1.75 m from the middle **5.** Provide your own answer **6.** Provide your own answer **7. A+B+C+D=E 8.** 40 J **9.** 5 m/s^2 **10.** wavelength **11a.** Hz, 1/s **b.** W/m^2, kg/s^3 **c.** m/s **12.** 2L, L, 2/3 L **13.** False **14.** .026 cm **15.** 2.46×10^7 **16.** Provide your own answer **17a.** 0 **b.** 20J **18a.** T, kg/C-s **b.** V, kg-m^2/C-s^2 **c.** F, C^2-s^2/kg-m^2 **d.** Ω, kg-m^2/C^2-s **e.** H, kg-m^2/s **19a.** 0 **b.** 2.16×10^7 N/C, outward **c.** 45 N/C, inward **20a.** 10 V/m, away from the plate **b.** 1.6×10^{-18} J **c.** 1.9×10^6 m/s **21a.** 0.545 F **b.** 4.545 F **c.** 3.5×10^{-4} Ω **22.** 2×10^{-5} N, away from the other wire **23.** Provide your own answer **24.** Provide your own answer **25a.** parallel **b.** series **26.** Out of the page **27.** False **28a.** 28.75° and 28.97° **b.** 0.05 cm **29.** True **30a.** 8.2×10^{-14} J **b.** 5×10^{-13} J **c.** 7.1 **31a.** 4.5×10^{-10} m **b.** 1.26×10^{-6} m **c.** 15.5×10^{-19} J **d.** 2.11×10^{-34} J-s **32.** 121.6 km (parallel to motion) and 76 km. **33.** Provide your own answer

Glossary

Note: Terms are followed, when appropriate, by the type of quantity—vector or scalar—and the conventional units of the quantity in the SI system. Common abbreviations are also noted.

absorption spectrum the specific wavelengths of light absorbed by a gas.

acceleration (vector; m/s^2) the rate of change of velocity.

adiabatic without the exchange of heat with the external universe.

alternating current (scalar; A) a type of electrical current that changes direction regularly in time.

alternating voltage an electrostatic potential that changes regularly in time.

ammeter a device used to measure electrical current.

ampere (A) the SI unit of measurement of current; equivalent to C/s.

amplitude (scalar; m) the maximal displacement of an object in simple harmonic motion; the peak-to-peak height of a wave.

angular acceleration (scalar; $radians/s^2$) the rate of change of angular velocity in time.

angular displacement (scalar; radians) the angle between the initial and final radii after a given time of an object moving in a circle.

angular momentum (vector; J-s) the cross product of the distance an object is from a point and its momentum with respect to this point.

angular velocity (scalar; radians/s) the rate of change of angular displacement in time.

antinodes points on a standing wave that have maximum displacement, due to the constructive interference of the component waves.

atomic mass the total number of nucleons in a nucleus.

atomic mass unit (amu) the unit of mass appropriate to the nucleus of an atom, equivalent to 1/12 of the mass of the carbon nucleus.

atomic number the number of protons in a nucleus.

Avogadro's number the number of objects or particles in a mole of substance, namely, 6.02×10^{23}.

beats the effect produced by interference of waves of slightly different frequency, producing a pattern of alternating intensity.

binding energy the energy difference between the mass energy of a nucleus and its separate constituent particles. Binding energy is released when nuclei come together during fusion.

blackbody an object that absorbs all radiation falling on it and radiates it perfectly.

Bohr radius the average radius of an electron orbiting a single proton, equal to 5.29×10^{-11} m.

Boltzmann's constant a fundamental constant, usually encountered in thermodynamics, with a value of 1.38×10^{-23} J/K.

British thermal unit (BTU) a unit of energy frequently used in engineering, equivalent to 252 calories or 1.054 kJ.

bulk modulus (scalar; N/m^2) the ratio of pressure to the resulting compressional strain.

buoyant force the upward force on an object placed in fluid.

calorie (cal) a unit of energy, defined as the energy required to raise one gram of water one degree Celsius, and equivalent to 4.184 J.

calorimeter a device used to measure the specific heat capacity of a substance.

capacitance (scalar; F) a measure of how much charge a capacitor can store.

capacitor an electrical device for storing separated charge and therefore storing electrostatic potential energy.

Carnot cycle an ideal series of states through which a heat engine may move, which maximizes the amount of heat converted into work. The Carnot cycle consists of two isothermal processes and two adiabatic processes.

centripetal acceleration (vector; m/s^2) the acceleration of an object in uniform circular motion which is pointed toward the centre of the circle.

centripetal force (vector; N) the force that holds an object in circular motion, pointed toward the centre of the circle.

charge (scalar; C) An intrinsic property of matter that causes it to produce an electric field and, when it is moving, a magnetic field, and to feel a force due to these fields as well.

circuit a closed loop arrangement of elements such as capacitors, resistors, inductors, and batteries, through which electrical current flows.

compressional strain (scalar) the fractional volume change of an object due to uniform pressure.

Compton scattering the deflection of electrons by photons.

concave mirror a mirror with an inwardly curved surface, such as the inner surface of a sphere.

conduction the transfer of heat through a stationary substance; also the movement of ions or electrons through a material.

conductor a material through which electrical charges can easily flow.

conservation of energy a fundamental law of physics which states that the energy of a system does not change unless an external force acts upon the system.

constructive interference the phenomenom in which waves present simultaneously at the same position add together to form a bigger wave.

convection the transport of heat due to the motion of heated substance, such as a gas.

converging lens a lens that causes parallel light rays to converge.

convex mirror a mirror with an outwardly curved surface, such as the outer surface of a sphere.

coulomb (C) the SI unit of measurement of charge.

critical angle the angle of incidence for a light ray, below which total internal reflection occurs.

current (scalar; A) the amount of electric charge passing a point per unit time.

daughter nucleus the nucleus remaining after a radioactive decay.

degrees (°) the gradations of a temperature scale; also a unit of measurement of an angle.

density (scalar; kg/m^3) mass per unit volume.

destructive interference the phenomenom in which waves present simultaneously at the same position add together to form a smaller wave, or momentarily cancel entirely.

dichroic having the property of naturally emitting light with only a single polarization.

dielectric constant (scalar) an intrinsic property of a substance that indicates the amount of charge induced in the substance when it is placed in an electric field.

diffraction the process of a wave spreading as it passes around an object or passes through a narrow opening.

direct current (scalar; A) a type of electrical current that flows in only one direction. Direct current flows between the terminals of a simple battery when they are connected to a circuit.

dispersion the effect of separating waves, especially light, into constituent wavelengths by passing through a medium with an index of refraction that varies with wavelengths.

displacement (vector; m) change in position of an object.

diverging lens a lens that causes parallel light rays to diverge.

Doppler effect the change in frequency of a wave produced by a moving source. Approaching sources have a higher frequency, and retreating sources have a lower frequency.

elastic collision a collision between two objects in which the kinetic energy of both is conserved.

elastic modulus (scalar; N/m^2) the ratio of stress to strain.

elasticity the property of an object to restore its shape after deformation.

electric field (vector; V/m) the force felt by a positive unit test charge in a region of space, due to the influence of other charges. Electric fields are produced by stationary and moving charges.

electric flux (scalar; V-m) the total sum of electric field vectors passing perpendicularly through a surface. According to Gauss's law, the electric flux through a closed gaussian surface is proportional to the total net charge contained within the surface.

electromotive force (emf; scalar; V) the electrostatic potential difference between the terminals of a circuit or battery when no current is flowing.

electron the negatively charged fundamental particles present in ordinary matter, surrounding the nucleus.

electroscope a simple device for indicating the presence of net electrical charge.

electrostatic potential (scalar; V) the amount of energy per unit positive charge required to move a charge between two points within an electric field.

emf see *electromotive force.*

emissivity (scalar) an intrinsic property of a material indicating how well it radiates heat.

energy (scalar; J) the ability to do work.

energy level diagram a diagram illustrating the discrete energies that an electron orbiting a nucleus may possess.

entropy (scalar; J/K) a fundamental thermodynamic quantity which measures how much heat energy is unavailable for conversion to work.

equipotential surface an array of positions within an electric field that all have the same electrostatic potential value. A charge may move along an equipotential surface without requiring or releasing energy.

equivalence principle the principle of general relativity which states that experiments conducted in an inertial frame in a gravitational field and experiments conducted in an accelerating reference frame will give the same results.

ether The medium on which lightwaves were once presumed to propogate. The luminous ether does not exist.

farad (F) the SI unit of measurement of capacitance; equivalent to A-s/V.

field lines a pictorial representation of an electric field or magnetic field or any other vector field.

focal length (scalar; m) the distance from the focal point of a lens or mirror to the surface of the lens or mirror.

focal point the point at which light rays from a mirror or lens intersect.

force (vector; N) a push or pull that causes an object to accelerate.

force diagram a diagram displaying all of the forces acting on an object.

forced vibrations the vibrations produced in an object connected to another vibrating object, with the effect of amplifying the vibrations of the first object.

free-body diagram another name for force diagram.

frequency (scalar; Hz) the number of complete cycles of simple harmonic motion per unit time; the reciprocal of period; the number of wave cycles passing a fixed point per unit time.

friction a retarding force between two objects that inhibits motion.

galvanometer a device used to measure electrical current.

gaussian surface an imaginary suface-enclosing charge used to calculate the electric field at points on the surface using Gauss's law.

general relativity the theory of mechanics that treats gravitational fields as equivalent to relative acceleration and introduces the notion that mass curves space and time.

generator a device that converts mechanical energy into electrical energy.

half-life the time required for half of a radioactive sample to decay.

heat capacity (scalar; J/K) the amount of heat energy required to change the temperature of an object by one degree.

heat of fusion (scalar; J) the heat that must be added per unit mass for a phase change of a substance between solid and liquid states. For a change from liquid to solid, the heat of fusion is released.

heat of vaporization (scalar; J) the heat that must be added per unit mass for a phase change of a substance between liquid and gas states. For a change from liquid to solid, the heat of vaporization is released.

henry (H) the SI unit of measurement of inductance, equivalent to V-s/A.

hertz (Hz) the SI unit of measurement of frequency; equivalent to s^{-1}.

ideal gas a collection of identical, infinitesimally small particles that interact only by elastic collisions.

impulse (vector; N-s) the product of the average force acting on an object and the time over which it acts.

index of refraction (scalar) an intrinsic property of a transparent substance, which measures the speed of light in the material compared to the speed of light in a vacuum.

inductance (scalar; H) a property of a circuit that is the proportionality constant between the rate of change of the current in that circuit and the emf that this changing current produces.

induction the process by which electromotive forces are generated in a circuit due to changes in a magnetic field.

inelastic collision a collision between objects in which the kinetic energy changes, for example, due to deformation or frictional loss.

inertial frame a set of coordinates that is not accelerating.

insulator a material through which electrical charges cannot flow.

intensity (scalar; W/m^2) the amount of energy carried by a wave across a unit area in a unit time.

isobaric at constant pressure.

isochoric at constant volume.

isothermal at constant temperature.

isotopes atoms with nuclei possessing the same number of protons but different numbers of neutrons.

joule (J) the SI unit of measurement of energy; equivalent to $kg\text{-}m^2/s^2$.

Kelvin (K) the SI unit of measurement of temperature.

kilogram (kg) the SI unit of measurement of mass.

kinetic energy (scalar; J) the energy of an object in motion.

kinetic friction friction that acts to resist motion of an object that is already moving.

kinetic theory of gases a model of an ideal gas that treats it as a collection of molecules undergoing motion according to Newton's laws of classical mechanics and predicts macroscopic quantities such as pressure and temperature in terms of molecular properties, such as the velocity of the molecules.

latent heat (scalar; J/kg) the heat required for a phase change of a substance per unit mass.

law of conservation of momentum a fundamental law of physics which states that the linear momentum of a system does not change unless an external force acts upon the system.

line spectrum the specific wavelengths of light absorbed and emitted by a gas. The line spectrum is different for each type of gas.

linear momentum (vector; kg-m/s^2) the product of mass and velocity.

lines of force see *field lines.*

longitudinal wave a wave in which points on the wave move parallel to the direction of wave propagation.

Lorentz contraction the effect that an observer moving with respect to a given object will find the object to be shortened compared with the measurement by an observer at rest relative to the object.

loudness (scalar; decibels) the intensity of a sound wave compared to a set standard, namely, the lowest threshold of human hearing.

luminous ether See ether.

magnetic field (vector; T) the force felt by a unit positive test charge moving through space, due to the influence of magnets or moving charges. Magnetic fields are produced by moving charges.

magnetic flux (scalar; T-m^2) the total sum of magnetic field vectors passing perpendicularly through a surface. According to Faraday's law, the rate of change of the magnetic flux through a surface is proportional to electromotive force generated on the closed loop containing the surface.

mass (scalar; kg) the intrinsic property of matter that causes it to resist acceleration.

mass spectrometer an instrument that separates ionized atoms or molecules based upon their ratio of charge to mass.

mole the measure of the number of objects or particles. One mole is 6.02×10^{23} objects (Avogadro's number).

moment of inertia (scalar; kg-m^2) the intrinsic property of an object that causes it to resist changes in rotational motion.

momentum (vector; kg-m/s) see *linear momentum* or *angular momentum.*

mutual inductance when a change in one circuit causes a change in a different circuit.

neutron the fundamental particle, possessing no charge, that is present in the nuclei of ordinary matter.

newton the unit of measurement for force; equivalent to kg-m/s^2.

nodal points points on a standing wave that do not move at all due to the destructive interference of the component waves.

normal force (vector; N) the force exerted by a surface on an object sitting upon it pointing perpendicular to the surface.

normal perpendicular.

north pole one end of a magnet; the end that attracts the south pole.

nuclear fission the process of a nucleus splitting into two nearly equal-sized nuclei.

nuclear fusion the process of two light nuclei coming together to form a heavier nucleus.

nucleon any particle present in the nucleus of an atom, that is, a proton or neutron.

nucleus the positively charged central collection of protons and neutrons in an atom.

ohm (Ω) the SI unit of measurement of resistance; equivalent to V/A.

opaque having the property of not allowing light to pass through.

parallel two lines or surfaces that never intersect. Also, for circuit elements, elements that are connected so that they have the same electrostatic potential difference across them.

parent nucleus the original nucleus during a radioactive decay.

pascal (Pa) the SI unit of measurement of pressure; equivalent to N/m^2.

period (scalar; s) the time of one complete cycle of simple harmonic motion; the time for one wave cycle to pass a fixed point.

permeability constant (μ_o) a fundamental constant of the universe, which is the proportionality constant in Ampere's law and has the value 1.26×10^{-6} T-m/A.

permittivity of free space (ε_o) a fundamental constant of the universe that is the proportionality constant in Coulomb's law.

phase the physical state of an object, for example, solid, liquid, or gas. Also, a property of waves indicating the displacement at the initial time.

photoelectric effect the emission of electrons by certain metals when the appropriate wavelength of light is shined on them.

photoelectrons the electrons emitted in the photoelectric effect.

photons discrete units of energy; the particles associated with electromagnetic radiation (light).

pitch (scalar; Hz) the frequency of a sound wave.

Planck's constant (h) a fundamental constant of the universe, appearing in quantum mechanics, with the value 6.626×10^{-34} J-s.

polarization a property of transverse waves, especially light waves, that indicates the orientation of the wave displacement with respect to a coordinate system.

polarizer a device that permits light of only one polarization to pass through.

potential energy (scalar; J) the energy of an object due to its position or internal structure.

power (scalar; W) the rate of change in energy, or the rate of doing work.

pressure (scalar; Pa) the ratio of force to the area to which the force is applied.

proton the fundamental particle, possessing positive charge, that is present in the nuclei of ordinary matter.

quanta discrete units of energy.

quantum mechanics the laws of physics that apply to objects on the atomic scale.

quantum number a number describing the quantum state of an object, such as the state of an electron orbiting a nucleus.

radians the SI unit of measurement of the size of an angle. A complete circle is 2π radians.

radiation the transport of heat in the form of electromagnetic waves.

radioactivity the process by which some nuclei spontaneously split and emit particles.

range the horizontal distance a projectile travels.

ray a straight line representation of the path of a light wave.

ray diagram a drawing of light rays used to analyze a set of optical devices, such as a lens or mirror.

RC circuit a circuit that contains a resistor and a capacitor in series with one another.

reactance (scalar; Ω) the apparent resistance of capacitors and inductors to alternating current.

real image an image produced in an optical device, such as a lens or mirror, that is formed by converging rays.

refraction the bending of light when it passes across the boundary between two media with different indices of refraction.

resistance (scalar; Ω) the proportionality constant between the applied electrostatic potential difference and the resulting current in a circuit.

resistivity (scalar; Ω–m) a measure of how well current flows through a material, per unit length and cross-sectional area.

resistor a circuit element that impedes the flow of current.

resonance a vibration produced in an object at its natural vibration frequency, or a multiple of it, due to a nearby object vibrating at this frequency.

rest mass (scalar; kg) the mass of an object, measured by an observer at rest with respect to the object.

restoring force the force exerted by a spring when it is stretched or compressed, which tends to restore the spring to its resting length.

resultant the sum of two or more vectors.

rigid body an object in which the relative distance between internal points does not change.

RL circuit a circuit that contains a resistor and an inductor in series with one another.

RLC circuit a circuit that contains a resistor, a capacitor, and an inductor in series with one another.

root mean square (rms) a way of averaging, equal to the square root of the average of the squares of a quantity.

rotational inertia (scalar; kg-m^2) see *moment of inertia.*

Rydberg constant (R) a constant found in the description of the line spectra of gases.

scalar a quantity with magnitude but without direction; expressed as a simple number.

self-inductance the generation of an electromotive force in a closed circuit due to a changing current in that circuit.

series for circuit elements, elements that are connected so that they have the same electrical current conducted through them.

shear modulus (scalar; N/m^2) the ratio of shear stress to shear strain.

shear strain (scalar) the ratio of the horizontal distance a sheared face moves to the height of the object.

shear stress (scalar; N/m^2) the ratio of the tangential force to the area of the face being stressed.

simple harmonic motion (SHM) the motion of an object with acceleration proportional to the displacement, resulting in repetitive motion.

simple pendulum a mass swinging on the end of a massless string under the influence of gravity.

solenoid a long straight coil of wire.

south pole one end of a magnet; the end that attracts the north pole.

special relativity the theory of mechanics for objects moving with velocities close to the speed of light.

specific heat capacity (scalar; J/kg-K) the heat capacity of a substance per unit mass.

spectrum the distribution of wavelengths or frequencies of electromagnetic radiation.

spring constant (scalar; N/m) the constant of proportionality between the applied force and the resulting change in length of a given spring.

standard pressure a convenient measure of pressure equivalent to 1 atmosphere or 1.01×10^5 pascal.

standard temperature zero degrees Celsius.

standard volume the volume of an ideal gas at standard temperature and standard pressure, namely, 22.4 litres.

standing wave a wave produced by the superposition of waves traveling in opposite directions, such that the sum is a wave pulse that does not move along the medium (see *nodal points, antinodes*).

static equilibrium the state of an object when all forces acting on it sum to zero.

static friction friction that resists an object's beginning motion.

stationary state a quantum state of an electron in which it does not emit radiation.

superposition principle the rule for adding waves at the same point together, which states that the resulting wave is the vector sum of all the independent waves.

sympathetic vibration see *resonance.*

temperature (scalar; K, °C) a measure of how hot or cold an object is with respect to a reference object.

tensile strain (scalar) the fractional change in length of an object due to an applied tensile force.

tensile stress (scalar; N/m^2) the ratio of tensile force to the cross-sectional area perpendicular to the force.

tesla (T) the SI unit of measurement of the magnitude of a magnetic field; equivalent to N/A-m.

thermal conductivity (scalar; J/m-K) an intrinsic property of a material indicating how well heat transfers through the material.

thermal contact contact between objects that allows them to influence each other's temperatures.

thermal equilibrium the state of objects in thermal contact when they no longer change each other's temperature.

thermal expansion the phenomenon that most substances increase in volume as their temperature increases.

thermodynamics the branch of physics concerned with very general properties of matter and energy. It also describes the macroscopic properties of matter in terms of the microscopic properties of its components.

thermometer a tool for measuring temperature.

threshold frequency the minimum frequency that light must have in order to cause photoelectrons to be emitted by a given metal.

time dilation the effect that time moves more slowly in an inertial frame moving with respect to a stationary one.

toroid a doughnut shaped coil of wire; a solenoid bent into a closed circle.

torque (vector; N-m) a push or pull that tends to cause an object to rotate about a fixed point; the rotational analog of force.

total internal reflection the process in which light traveling from a material with a higher index of refraction to a material with a lower index of refraction is reflected at the boundary and no light actually crosses the boundary.

transformer a device that is used to pass an alternating voltage from one circuit to another. In the process, the voltage may be increased or decreased.

transmutation the process of one nucleus turning into another through radioactivity.

transparent having the property of allowing light to pass through.

transverse wave a wave pulse in which points on the wave move perpendicular to the direction of wave propagation.

triple point of water the temperature of water at which ice, liquid water, and water vapour coexist at thermal equilibrium, defined as 0.01°C or 273.15 K.

uncertainty principle the statement that, due to the laws of quantum mechanics, it is impossible to simultaneously exactly measure a particle's position and momentum or to exactly measure a particle's energy for a finite amount of time.

universal gravitational constant (G) proportionality constant in Newton's law of universal gravitation, a fundamental constant of the universe, with the value 6.67×10^{-11} N-m^2/kg.

vector a quantity with magnitude and direction.

vector component the projection of a vector upon a given coordinate axis.

velocity (vector; m/s^2) the rate of change of position in time.

virtual image an image produced in an optical device, such as a lens or mirror, that is formed by the extension of diverging rays.

volt (V) the SI unit of measurement of electrostatic potential; equivalent to J/C.

voltmeter a device used to measure electrostatic potential differences between two points.

watt (W) the SI unit of measurement of power; equivalent to J/s.

wavelength the distance between identical points in a wave cycle.

weight (vector; N) the product of mass and gravitational force.

work (scalar; J) the product of the force exerted on an object and the distance through which the object moves as a result. Work results in a change in energy.

work function the energy required to release an electron from a metal due to the photoelectric effect.

Young's modulus (scalar; N/m^2) the ratio of tensile stress to tensile strain.

Notes

Notes

Notes

Notes

Notes

Notes

Notes

Notes